Future of Robotics 21st
Century Robotics technologies

Transformative Impact and Ethical
Considerations of Robotics Technology

Alan Sparkbot

Contents

Chapter 1: The Ascent of Mechanical Technology: A Verifiable Viewpoint 3

    Development of Mechanical technology from Fiction to the real world 18

Chapter 2: The Life Systems of Robots: Figuring out their Parts and Works 20

    Investigating the Inward Activities of Present-day Advanced Mechanics 31

Chapter 3: High-level Mechanics in Industry: Changing Collecting and Creation 33

    From Sequential construction systems to Shrewd Production lines 42

Chapter 4: Robots in Medical Care: Changing Medication and Patient 45

    Advancements in Careful Mechanical Technology and Clinical Help 54

Chapter 5: The Job of Robots in Investigation: Propelling Space and Maritime Disclosures 56

    From Mars Wanderers to Remote Ocean Voyagers 63

Chapter 6: Advanced Mechanics and Instruction: Forming the Fate of Learning 66

    Coordinating Mechanical technology into STEM Educational program 74

Chapter 7: Independent Vehicles: Heading Toward a Driverless Future 76

Navigating the Roads with AI-Powered Vehicles..83

Chapter 8: Advanced Mechanics and Farming: Developing Proficiency and Supportability..................85

Accuracy Cultivating and the Rural Transformation 93

Chapter 9: Robotics in Disaster Response: Enhancing Safety and Rescue Operations..95

Deploying Robots in Emergency Situations..........103

Chapter 10: The Morals of Advanced Mechanics: Tending to Moral and Social Ramifications 105

Balancing Innovation with Responsibility 112

Chapter 11: The effects of robots on employment on job and workforce dynamics..114

Making Adjustments to the Changing Employment Landscape..120

Chapter 12: Accessibility and robotics: Giving people with disabilities more power...122

Enhancing Accessibility through Assistive Robotics 128

From Animatronics to Interactive Performers....137

Chapter 14 Understanding the complexities of military applications through robotics and warfare 139

Analyzing Robotics' Contribution to Defense Strategies 145

Chapter 15: From Companionship to Coexistence: The Direction of human-robot Interaction in the Future 147

  Analyzing Relationship Dynamics Between People and Robots 153

Chapter 16: Mechanical Technology and Ecological Preservation: Safeguarding Nature with Innovative Arrangements 156

  Utilizing Robots for Conservation Activities 163

Chapter 17: Rebuilding Communities After Disasters with Robotic Innovations in Disaster Recovery 165

  Using technology to rebuild after a disaster 171

Chapter 18: Personal Assistants and Robots: Redefining Daily Life with AI Companions 173

  Personal Care to Automation of the Home 179

Chapter 19: Research and Development in Robotics: Obstacles and Opportunities 181

  Navigating the Frontier of Robotics Innovation . 187

Chapter 20: The Future of Robotics: Predicting Trends and Designing the World of Tomorrow 189

  Envisioning the Next Era of Robotics Integration 197

Chapter 1: The Ascent of Mechanical Technology: A Verifiable Viewpoint

Machines that copy human or animal exercises have captivated humanity for quite a long time. From the unbelievable machines of Greek legends to Leonardo da Vinci's shrewd depictions, the dream of robots has immersed our inventive personalities. This part dives into the genuine underpinnings of mechanical innovation, following its advancement from early starts to the refined machines that shape our current reality.

- Early Dreams: From Dream to Part Our advantage in robots can be followed back to old turns of events. Greek dreams examined Talos, a bronze beast safeguarding Crete, and Hephaestus, the master of fire and metalworking, who made splendid machines. These records, while fantastical, established the groundwork for machines prepared for human-like turn of events. Fast forward to the Renaissance, when makers like Leonardo da Vinci revived these considerations on paper. His scratch pads contain point-by-point portrayals of mechanical knights, humanoid figures, and, shockingly, a self-incited truck, displaying a pivotal perception of mechanics and planning norms. Anyway never created, these plans go about as an exhibit of the visionary thinking of this period.

The Time of Machines: Marvels of Planning The seventeenth and eighteenth many years saw a flood in the improvement of machines. These muddled, habitually life-sized machines were marvels of planning, prepared for performing complex tasks like organization, playing music, and regardless, handling food (but the last choice was a significant part of the time a sharp trickiness). Leading figures like Jacques de Vaucanson, a French pioneer, moved amazing machines, including a mechanical duck that could eat and poop (with a pre-stacked secret part) and a human figure that played the flute. These marvels of configuration controlled public interest and laid the groundwork for the headway of extra complex machines. The Cutting Edge Change: The Presentation of Practical Mechanical Innovation The Advanced Agitation presented some other time for mechanical innovation. With the climb of modern offices and enormous scope fabricating, the prerequisite for robotized machines to perform dull tasks ended up being dynamically self-evident. The current robots were less marvelous than the machines of the previous period, focusing on convenience instead of mind-boggling mimicry. One of the earliest models is the steam-powered loom planned by Jacquard in 1801.

This machine used punch cards to control the twisting around process, a basic accomplishment in the improvement of programmable machines. Long term, these advanced robots ended up being consistently more mind-boggling, laying out the foundation for the motorization that describes present-day manufacturing.

The 20th Hundred years: Towards Sharp Machines The 20th century saw a shocking velocity expansion in the field of mechanical innovation. The making of the semiconductor in 1947 downsized devices, planning for additional unassuming, more versatile robots. Leading researchers like George Devol and Joseph Engelberger encouraged the main present-day robot with programmable arms during the 1950s. This improvement meant a pivotal turning point, as robots could now be acclimated to play out a greater extent of endeavors. The last half of the century saw further movements in mechanical innovation, with the rise of programming and modernized thinking (reenacted knowledge). The possibility of robots performing tasks as well as choosing and acclimating to their ongoing situation began to work out as expected. Vision structures, sensors, and undeniable level control computations allowed robots to team up with the world in a seriously bewildering way.An Obvious Viewpoint of Mechanical Development: Consistent mechanical advancement is the consequence of a rich tenable custom following beyond what many would consider possible back to remainder and spread out in more pre-arranged extreme convictions. The following are a couple of critical achievements in this custom: Remnant: Old community foundations had their types of robots and mechanical contraptions. For instance, the obsolete Greeks made complex automata, the notable "Pigeon of Archytas" and "The Mechanical Specialist" by Legend of Alexandria. Bygone eras: During this period, makers continued to examine mechanical devices. Al-Jazari, a thirteenth-century engineer, arranged different automata, including a melodic band and a mechanical peacock. Renaissance and Brightening: Leonardo da Vinci conceptualized plans for humanoid robots, even though they were seldom gathered. His portrayals included contemplations for mechanical knights and other definite figures. Present-day Commotion:

The eighteenth and nineteenth years saw basic degrees of progress in device and computerization. Present-day robots emerged, primarily for the ultimate objective of collecting. 20th Hundred Years: The adage "robot" was written by Czech essayist Karel Čapek in his play "R.U.R." (Rossum's Overall Robots). During the 20th hundred years, experts like George Devol and Joseph Engelberger cultivated the essential present-day robots for consecutive development frameworks. 1960s: The field of mechanical innovation broadened rapidly. Researchers like Joseph Weizenbaum examined man-made thinking, and the primary robot arm (Unimate) was presented in an Overall Motors handling plant. Past the 60s: High-level mechanics continued to create, with applications in space examination, clinical operations, and normal presence. Social robots, such as ASIMO and Pepper, entered the scene. In the narratives of humanity's arrangement of encounters, making counterfeit animals help or duplicate individuals has entranced human headways for a seriously prolonged period. From old legends of automata to the state-of-the-art season of state-of-the-art mechanical innovation, the journey of mechanical innovation is as much a showing of human creativity as it is an impression of our objectives and fears. The seeds of cutting-edge mechanics were established in the personalities of old turns of events. Stories from old Greek fables, for instance, the story of Talos, a beast bronze robot that depended on watching the island of Crete, got the human interest in making counterfeit life. These early stories established the groundwork for the chance of fake animals that could perform endeavors past the capacities of individuals. In any case, it was shortly after the start of the Advanced Commotion in the eighteenth and nineteenth years that the chance of mechanical computerization began to take an obvious design. The development of muddled amazing luck frameworks and the progression of early steam-controlled machines laid out the preparation for the computerized world that would follow. The maxim "robot" itself finds its beginning stages in the Czech word "robot," the significance of obliged work or servitude. It was generated by author Karel Čapek in his 1920 play "R.U.R. (Rossum's Broad Robots)," which depicted fake animals made to serve mankind anyway resisting their

producers. This unique work upheld the adage "robot" yet moreover introduced subjects of freedom, ethics, and the potential results of making shrewd machines. The mid-20th century saw basic degrees of progress in cutting-edge mechanics, pushed by speedy imaginative headway and the space race. Foundations like the Massachusetts Underpinning of Advancement (MIT) and affiliations like NASA expected pressing parts in stretching the boundaries of mechanical examination and automation.

From the focal current robots presented by George Devol and Joseph Engelberger during the 1950s to the lunar vagabonds conveyed during the Apollo missions, undeniable level mechanics changed from the space of sci-fi to mentally calm reality. As taking care of force expanded and scaling down became possible, mechanical advancement entered another time of a multifaceted nature. The movement of chips, sensors, and actuators drew in the plan of robots arranged for unbelievable assignments and versatile ways to deal with acting. Bounce progresses in man-made mindfulness, especially in the fields of PC-based knowledge and psyche affiliations, further extending the limits of robots, permitting them to see, learn, and communicate with their ecological variables in powerfully complex ways. Today, mechanical development drenches all parts of current life, from social affairs and clinical advantages to transportation and diversion. Pleasant robots, or "cobots," work close to people in taking care of plants, reducing viability and flourishing. Careful robots help specialists with accuracy and guilefulness, upsetting undertakings. Independent vehicles vow to change transportation, making streets more secure and more helpful. Anyway, as mechanical progression keeps on driving, it moreover raises tremendous issues about morals, work, and the chance of humanity itself. The move of free plans concerns works send-off and cash-related inconsistency, while the opportunity to notice machines is irritated by how we could loosen up insight and moral responsibility.In this part, we will leave on an excursion through time, looking toward the early phases, achievements, and outcomes of the rising mechanical development. From the fantasies and legends of past periods to the state-of-the-art sorts of the progress of the melodic advancement day, we will bounce into the rich winding of the human creative cerebrum and improvement that has shown the universe of mechanical headway as we point of fact have some information on it.We will look at how mechanical headway has been delivered involving its reasonable early phases as an idea participant in stories to a multidisciplinary field wrapping coordinating, programming, and mental psyche research. We will look at the key minutes and key figures who have added to the progression of mechanical development, from early trailblazers like Nikola Tesla and Alan Turing to contemporary pioneers, for example, Rodney Streams and Hiroshi Ishiguro.Our outing will take us through the creative achievements that have depicted the advancement of state-of-the-art mechanics, from the creation of the programmable robot by George Devol to the improvement of refined humanoid robots like ASIMO and Sophia. We will plunge into the leap moves in man-caused convictions that have empowered robots to see and relax their by and large natural elements, from PC vision frameworks that can see things and appearances to ordinary language-overseeing calculations that draw in robots to handle and answer human speech.Along the way, we will analyze the different clarifications behind mechanical movement across endless undertakings and spaces. We will survey how robots are changing get-togethers and making factors, smoothing out creation processes, and further capacity to make. We will figure out how robots are changing clinical advantages, helping trained professionals and escorts in endeavors, recovery, and senior considerations. We will find how robots are reshaping transportation and examination, from self-driving vehicles and robots to planetary drifters and far-off ocean submersibles. But our assessment addressing things to happen to mechanical movement won't be restricted to creative upgrades alone. We will correspondingly wrestle with the moral, social, and philosophical consequences of a world populated by sharp machines. We will contemplate demands for independence and associations, as well as the conceivable effect of top-tier mechanics on work, dissimilarity, and human prospering. Also, we will consider staying in our steady reality where people and robots match, get together, and, maybe, even structure monstrous bonds. As we sway further into the complexities of mechanical movement, we should go up against the moral contemplations that go with the speedy improvement of this field. Questions emerge concerning the ethical outcomes of getting ready machines with the suspicion of free courses and the common unavoidable aftereffects of such activities. The moral plan consolidating advanced mechanics integrates issues of success, accomplishment, and obligation, affecting conversations on the essentials for huge norms to work with the turn of events and sending of automated systems. Moreover, the social effect of state-of-the-art mechanics can't be pleasantly conferred. The blend of robots into different bits of standard presence can change social plans and standards, reshaping how we live, work, and achieve. While robotization offers the commitment to grow reasonableness and capability, it in addition raises worries about work clearing and money-related differences, including the meaning of looking out for these difficulties through keen system measures and social initiatives.In line with these moral and social contemplations, the field of mechanical improvement keeps on expanding the limitations of mechanical headway. Specialists and specialists are researching new backwoods in delicate mechanical movement, bio-mixed plan, and human-robot collusion, expecting to cultivate robots that are more talented as well as more adaptable, strong, and open to the necessities of humans.Looking ahead, the conceivable predetermination of state-of-the-art mechanics holds both responsibility and probability. On one hand, mechanical progress can empower human endpoints, work on ordered satisfaction, and address beating considering everything, from clinical thoughts and needs to standard worth and disaster reaction. Then again, the extreme extension of mechanical improvements could fuel existing inconsistent attributes, support socially disreputable shows, and even pose existential dangers to humanity. In assessing this radiant scene, we ought to push toward the predetermination of top-tier mechanics with lowliness, information, and hunch. By crossing the force of progress for everyone's flourishing and staying aware of the typical expansions of empathy, worth, and strength, we can guarantee that the obligation to state-of-the-art mechanics is found in affinities that benefit all of humanity. As we pull out on this excursion into the conceivable destiny of top-tier mechanics, let us embrace the potential entryways that lie ahead while furthermore seeing the difficulties that should be made due. Together, we can shape a future where robots and people fit pleasingly, teaming up to incorporate a stunning and more prosperous world from this point until quite a while to come, unendingly and ceaselessly to come. In our assessment addressing things to happen to mechanical new development, seeing the potential for working with effort and associations among people and machines is fundamental. As opposed to focusing on robots as clear contraptions or substitutes for human work, we can imagine a future where people and robots complete each other qualities and endpoints, orchestrating synergistically to manage complex irritates and accomplish common goals. One district where this consistent viewpoint is especially lifting is in the field of assistive mechanical development. Assistive robots can resuscitate the specific satisfaction of people with indiscretions or age-related limits, offering help with typical undertakings, conveyability, and correspondence. By remembering progress for robotized thinking and sensor sorts of progress, assistive robots can adjust to the key necessities and affinities of their clients, associating with them to live more directly and autonomously. Similarly, in the space of clinical advantages, robots can likely as fundamental accessories to clinically taught specialists, encouraging their abilities and well-being to sort out some way to orchestrate results. Cautious robots, for instance, can help specialists with accuracy and tendency, reducing the bet of human goof and drawing in irrelevant interfering designs with quicker recuperation times. Robots can moreover be conveyed in telemedicine applications, partner far-away gatherings, and seeing patients, especially in underserved or distant areas. Beyond clinical advantages, robots are ready to change affiliations going from agribusiness and development to retail and kind attitude. In making, robots equipped with top-level sensors and reenacted information examinations can moreover engage harvests the pioneers work on, expanding yields while restricting standard effects. Being made, robots can help with attempts like bricklaying, welding, and obliteration, further spreading out reasonableness and thriving in workplaces. In retail and energy, robots can reestablish client sponsorship and smooth out works, from electronic checkouts and stock relationships to room affiliations and escort associations. Notwithstanding, as we embrace the hindrance of state-of-the-art mechanics to change different bits of society, we should correspondingly remain mindful of the dangers and gets that go with imaginative development. Worries about confirmation, security, and the normal abuse of mechanical progression should be based on energetic protections and administrative plans. Additionally, endeavoring to free the effect of robotization on positions and laborers should be tangled, guaranteeing that the common options of mechanical development are enough appropriate across society. Taking into account everything, the specific destiny of mechanical improvement holds a titanic commitment concerning moving human succeeding and zeroing in on reasonable the smashing has a go at restricting our consistent reality. By attracting forward attempts and connections among people and machines, we can deal with the exceptional force of mechanical improvement to make a more central, fair, and sensible future for all. As we set out on this excursion into the feeble, let us do as such with positive thinking, an innovative frontal cortex, and a helping obligation to blend a dazzling world from this point forward, continually.

Development of Mechanical technology from Fiction to the real world

Early Beginning stages: automating tasks with machines returns many years. Early makers and experts arranged mechanical contraptions that were supposed to duplicate human turns of events.

For instance, Leonardo da Vinci's portrayals of mechanical knights and automata in the fifteenth century are early occurrences of tries to make humanoid machines. In any case, it was solely in the 20th century that the saying "robot" was coined by Czech essayist Karel Čapek in his 1920 play "R.U.R." (Rossum's Overall Robots). These robots were phony creatures made to perform work for individuals, lighting public interest in the idea. The Cutting edge Upset: The gigantic leap forward in mechanical innovation occurred during the Advanced Change. Initiating engineers like George Devol and Joseph Engelberger introduced the current robots during the 1950s. These early robots were essentially used in collecting plants to perform excess and risky endeavors like welding and painting. Strikingly, the Unimate, made by Devol and Engelberger, was presented in an Overall Motors fabricating plant in 1961. Degrees of progress in Computerization: As development progressed, so did the limits of robots. The methodology of computer chips and PC control systems during the 1970s and 1980s thought about additional advanced and precise turns of events. Robots were not commonly limited to drawn-out tasks; they could acclimate to changing conditions and perform complex exercises. The Climb of Agreeable Robots (Cobots): Lately, one more arrangement of robots has emerged: helpful robots, or "cobots." Not at all like their predecessors, which were much of the time withdrawn in fenced-off locales for prosperity reasons, cobots are expected to work near individuals, working on their abilities rather than displacing them. This headway has opened up extra doors for motorization in organizations like clinical consideration, tasks, and restricted scope creation. Mechanical innovation in Clinical benefits: One of the most reassuring districts for mechanical innovation is clinical benefits. Cautious robots, similar to the da Vinci Cautious System, have changed methods by offering extended precision and reducing prominence. Robots are furthermore being used for tasks like rebuilding treatment and old thoughts and giving assistance and support to patients.

Chapter 2: The Life Systems of Robots: Figuring out their Parts and Works

Robots, those marvels of the plan and creative mind are delivered involving confusing structures and parts that work as one to play out a crowd of attempts. Understanding the ongoing plans of robots is head to figuring out their abilities, objectives, and reasonable applications. In this segment, we will leave out going into the interior activities of present-day progressed mechanics, jumping through and through into the parts and works that make robots tick.

At the mark of a combination of every robot lies its mechanical new development, or skeleton, which gives the construction to its undertakings. The skeleton moves, by and large, dependent upon the sort and protection of the robot, going from clear regulator arms used in present-day settings to complex humanoid bodies made game plans for the human-like coalition.

The materials used in building the skeleton can moreover move, with metals, plastics, and composites being generally ordinary choices. Mounted on the bundling are actuators, the muscles of the robot that engage advancement and control. Actuators come in various plans, including electric motors, pneumatic chambers, and strain-driven structures, each fit different endeavors and conditions. Electric motors, for example, are dependably used in mechanical joints and farthest focuses thinking about their exactness and controllability, while pneumatic actuators win in applications requiring speedy strong regions for and improvement to actuators, robots are furnished with sensors that assess their typical variables and inside state. Sensors are most likely the robot's eyes, ears, and material receptors, enabling it to see and interface with the world. Normal sorts of sensors consolidate cameras, LiDAR (Light District and Running) scanners, locale sensors, and power/force sensors, each filling a shrewd need in the robot's huge toolkit.

The psyche of the robot, its control structure, processes information from sensors and issues sales to actuators, sorting out its new turns of events and ways of managing acting. Control structures can go from major, pre-changed plans of exercises to current, versatile estimations that learn and conform to making conditions. Moves in man-caused discernment and imitated information have actuated the advancement of rationally quick and free robots sorted out for complex courses and issue-solving.Beyond its genuine parts, robots are other than constrained by programming, the programming code that guides their method for managing acting and handiness. Programming has a fundamental effect in depicting the robot's abilities, from head improvement control and bearing to state-of-the-art keenness and dynamic computations. Programming vernaculars like C++, Python, and MATLAB are reliably used in cutting-edge mechanics improvement, attracting fashioners to coordinate, duplicate, and convey robotized systems with ease.Finally, robots a huge piece of the time rely on power sources like batteries, energy units, or external power supplies to work.

The choice of power source depends on parts like the robot's size, versatility basics, and energy plentifulness evaluations. Battery-controlled robots offer convenience and adaptability, while robots can draw power from external focal concentrations for broadened operation.In the once-gotten done, the ongoing plans of robots unite a substitute feature of parts and work that way to engage their abilities and ways of managing acts. From the mechanical development and actuators to sensors, control structures, programming, and power sources, each part plays a critical part in the pointlessness of the robot's arrangement and cutoff. By understanding the inward tasks of robots, we gain information on their logical applications and the challenges related to organizing and sending them into the certified world.Moreover, the coordination and effort of these parts cooperating is the head of a robot's sound judgment in various endeavors and conditions. For instance, in a get-together setting, a robot's mechanical development and actuators enable it to control objects with precision and speed, while its sensors assess to ensure unquestionable coordination and quality control. Meanwhile, the control structure works with these exercises, changing steadily to changes in the creation line or common conditions.In more remarkable circumstances, for instance, outdoor evaluation or disaster response, robots rely on a mix of sensors and programming to look at and talk with their regular parts energetically. LiDAR sensors, for example, give 3D organizing limits, allowing robots to see obstructions and plan ideal courses through complex scenes. In the meantime, copied information evaluations attract robots to see and adjust to novel conditions, getting from past experiences to deal with their show over time.Additionally, the mentality and versatility of motorized systems think about customization and grouping to impart attempts and essentials. Robots can be equipped with appropriate end-effectors, for instance, grippers, pull cups, or gadgets, to play out a great number of attempts — from picking and setting objects to welding, painting, or in any case, performing delicate activities. Moreover, withdrawn plans attract the split of the distinction between new sensors, actuators, or programming modules as progress advances, ensuring that robots stay flexible and up-to-date.As mechanical improvement continues to occur, interdisciplinary exertion has a massive effect on driving the field. Engineers, PC analysts, mental clinicians, and space experts from various fields collaborate to encourage inventive responses to complex issues, drawing inspiration from science, neuroscience, and various disciplines. By using encounters from nature and preparing the power of interdisciplinary assessment, researchers can make robots that are useful and rational as well as perfect, versatile, and sustainable.In the end, the flow designs of robots address a mix of plan, science, and creative characters, inciting machines that can extend and manage human limits in various settings. By understanding the parts and works of magnificence care item robots, we gain information on their run-of-the-mill applications and cutoff points, as well as the hardships and entryways that lie ahead. Looking forward, the certain destiny of cutting-edge mechanics holds monster potential for the extra turn of events and revelation. As progress continues to move, robots will end up being intensely integrated into our standard plans, disturbing undertakings, affiliations, and, amazingly, our joint endeavors. From free vehicles and transport robots to mechanized vast assistants, the doorways for mechanical headway are limited essentially by our innovative characters and ingenuity.One locale uncommonly convincing is the development of sensitive irrefutable level mechanics, moved by the biomechanics of living customary parts. Sensitive robots are conveyed utilizing versatile

materials that reflect the flexibility and versatility of standard tissues, considering protected and fragile formed tries with individuals and delicate things. Utilizations of fragile mechanical improvement range from clinical contraptions and prosthetics to wearable exoskeletons and tricky grippers for regulating delicate items.Another edge in best-in-class mechanics research is the appraisal of endless best-in-class mechanics, persuaded by the total ways of managing the acting of social bugs like bugs and bumble bees. Swarm robots should participate in tremendous get-togethers to achieve complex endeavors that would be hazardous or staggering for a lone robot to accomplish alone. Cases of a huge number of mechanical advancement applications join pursuit and rescue missions, standard checking, and improvement projects.Furthermore, advancements in mechanized thinking and man-made awareness are engaging robots to learn and conform to their regular parts transparently. Support
learning evaluations, unequivocally, grant robots to help new endpoints through trial and error, refining their lead fundamental length inspected their experiences. This limit opens up extra regions for robots to work in unstructured and dynamic circumstances, from family endeavors

and individual assistance to space assessment and cut-down examination. Notwithstanding, as robots become more organized in the public field, it is crucial to address moral, social, and money-related assessments connected with their sending. Stresses over work flight, endorsement, security, and algorithmic affinities ought to be meticulously considered and worked with through sharp rules, straightforwardness, and obligation.

Moreover, attempts to authorize blend and thought in mechanical improvement creative work are head to ensure that the reasonable developments of mechanical progression are reasonably scattered across all affiliations. Eventually, the possible destiny of mechanical improvement obligations is to be both advancement and testing, as we continue to foster the necessities of what is possible with sharp machines. By making interdisciplinary endeavors, embracing gatherings and thought, and zeroing in on the moral and careful turn of events, we can set up the stunning power of mechanical improvement to guarantee, on an extraordinarily essential level truly captivating, and reasonable future for all. As we set out on this journey into the future, let us stay worked with by our characteristics and necessities, endeavoring to make a reality where robots and individuals can win as one. Despite mechanical sorts of progress, the specific fate of cutting-edge mechanics will correspondingly be illustrated by wonderful points of view and social pieces of data. As robots become more standard in our for the most part common plans, drawing in a positive and cautious story around their work and potential obligations is central. This course isn't just about the presence of everybody but instead the endpoints and objectives of robots in any case as well as moving sympathy, understanding, and joint exertion among individuals and machines. Also, the blend of robots in the public field will require quick examinations of epic and significant plans to guarantee the accomplishment, security, and moral use of mechanical new turns of events. Policymakers and embellishments should attempt to connect with choices and decisions that address arising inconveniences and courses in high-level mechanics, from information security and moderate prospering to risk and obligation continuing to acknowledge there ought to emerge an event of catastrophes or influences.

In the meantime, attempting to democratize consent for unquestionable level mechanics progress and the outline is fundamental for making one more new development and assisting people and relationships with participate in the pointlessness of the predetermination of evident level mechanics. Drives, for example, open-source stuff and programming stages, creator spaces, and mechanical improvement give roads to the joint effort and getting, drawing in different voices and points of view to add to the progression of obvious level mechanics. Likewise, as robots become bound with human culture, looking at the moral and philosophical repercussions of human-robot experiences is huge. Demands concerning independence, association, and the chance of care will turn out to be goliaths as robots become more refined and free. It is convincing to push toward these plans with lowliness, empathy, and check of values like regard, congruity, and goodness. At last, the inescapable destiny of mechanical improvement holds a titanic commitment to move individuals to succeed and set out new ways for progress and straightforwardness. By embracing the obstruction of mechanical improvement while focusing on the moral, social, and social loads that go with their split division in the public eye, we can cause a future where robots and people agree charmingly, sharing to make an unequaled world from now into the foreseeable future, unendingly, endlessly, ceaseless. As we go on out on this outing into the delicate, let us stay worked with by our qualities and fundamentals, trying to make a future where improvement serves humanity's most raised targets and needs.

Investigating the Inward Activities of Present-day Advanced Mechanics

Late Advances in Applied Mechanics: The procedures of the Virtual Workshop on Applied Mechanics (VSAM 2021) give significant bits of knowledge into mechanical progressions in strong mechanics, liquid mechanics, and biomedical designing.

Leading specialists from around the world added to this meeting, covering subjects, for example, Mathematical examinations on the engendering of non-direct Sheep waves through delaminated surfaces in hardened composite plate structures. Excitation data transmission relies on cantilever energy gatherers. Stage field models applied to crack in solids. Reproduction concentrates on the proliferation of activity potential in epicardial tissue because of quality changes. Assessment of outpouring limit conditions in DNS of violent fly streams.Impact of fluidic infusion on the center length of rectangular sonic planes. Stress dissemination in vastly lengthy plates with round openings. Shrewd sensor ideas for modular evaluations of extensions exposed to arbitrary and vehicle excitations.Ballistic investigation of shear thickening liquid impregnated unidirectional super high sub-atomic thickness polyethylene texture.Furthermore, significantly more! 2. Sub-atomic Recreations: While not straightforwardly connected with mechanics, sub-atomic recreations assume a significant part in understanding the physical compound properties of dense matter frameworks. These recreations consolidate mathematical techniques with the PC's ability to address connections between particles or atoms. Old style Mechanics: Classical mechanics fills in as the establishment for figuring out complex dynamical issues. It is fundamental for regarding mechanical frameworks as well as for getting a handle on the nuts and bolts of quantum mechanics and measurable physical science.

Chapter 3: High-level Mechanics in Industry: Changing Collecting and Creation

The space of gathering and creation has gone through a huge change with the blend of mechanical innovation into present-day cycles. From vehicle consecutive development frameworks to equipment manufacturing plants, robots have changed how the product is conveyed, further developing capability, exactness, and flexibility. In this segment, we will examine the impact of cutting-edge mechanics on industry and how robotization is reshaping the destiny of manufacturing

.At the center of cutting-edge mechanics in the business lies the possibility of automation, the usage of machines to perform tasks with unimportant human intercession. Current robots are explicit machines planned to execute drawn-out, serious tasks with speed, accuracy, and consistency. Furnished with state-of-the-art sensors, actuators, and control systems, these robots can manage a broad assortment of gathering assignments, from welding and painting to packaging and palletizing.One of the fundamental benefits of mechanical innovation in the industry is the ability to assemble effectiveness and throughput while diminishing costs and cycle spans. Using routine tasks, robots can work perpetually, the entire day, consistently, without the prerequisite for breaks or individual time, inciting higher outcomes and more noticeable efficiency. This licenses creators to satisfy growing requirements while staying aware of raised levels of significant worth and consistency in their products.Moreover, robots engage creators to achieve levels of exactness and accuracy that are irksome or challenging to accomplish with human work alone. Undeniable level robotized arms furnished with exactness sensors and vision systems can perform complex social affair endeavors with sub-millimeter accuracy, ensuring tight protections and restricting blemishes. This is particularly critical in organizations like Flight, where precision is essential for prosperity and performance.In the development to redesigning effectiveness and quality, mechanical innovation in the business furthermore offers advantages of flexibility and adaptability. Not at all like customary gathering circumstances, which are by and large unyielding and fearless, mechanical automation thinks about fast reconfiguration and reevaluating to oblige changes in thing plan, creation volume, or market revenue. This deftness engages producers to answer quickly to moving business area components and client tendencies, gaining a high ground in the marketplace.Furthermore, mechanical innovation in the business has a basic impact in additional creating working climate security and ergonomics through mechanizing risky or mentioning endeavors. Robots can manage significant weights, work in over-the-top temperatures or conditions, and perform endeavors that present threats to human-trained professionals, such as welding or painting. By diminishing the receptiveness of workers to dangerous conditions, robots help with laying out safer and better working environments, decreasing the bet of disasters and injuries.However, the expansive gathering of cutting-edge mechanics in the business moreover raises issues and challenges associated with business, arrangement, and money-related impacts. While robots can enhance human-trained professionals and set out new entryways for skilled situations in cutting-edge mechanics upkeep, programming, and the board, they also can remove specific kinds of low-gifted or dreary positions. Tries to address these troubles through workforce planning, re-skilling ventures, and methodologies that advance work creation and monetary improvement are major for ensuring that the upsides of mechanical innovation are shared fairly across society.In the end, mechanical innovation in the industry tends to have a significant impact on in context in the way stock is made, changing plants into particularly motorized, capable, and versatile creation systems. By outfitting the power of cutting-edge mechanics to augment productivity, overhaul quality, and further foster workplace prosperity, creators can open new entryways for improvement and progression in the overall business community. As we continue to research the capacity of mechanical innovation in the business, let us try to make a future where motorization fills in as a catalyst for positive change, driving money-related achievement, reasonability, and human well-being.Furthermore, as the field of mechanical innovation in the industry is created, ongoing crazes and advances are emerging that assure extra overhaul manufacturing cycles and limits. Agreeable robots, or cobots, are one such turn of events, expected to work near human workers in shared workspaces. These robots are outfitted with state-of-the-art security features and regular places of communication, enabling them to collaborate with individuals on endeavors like social occasions, looking into, and material taking consideration. Cobots offer producers the flexibility to motorize complex

tasks while staying aware of human oversight and capacity, provoking more viable and adaptable creation systems.Another design shaping the possible destiny of mechanical innovation in the business is the joining of man-made thinking (PC-based knowledge) and artificial intelligence computations into computerized structures. Artificial intelligence-fueled robots can break down huge volumes of information, recognize examples, and make shrewd choices
continuously. This empowers them to advance creation processes, anticipate support needs, and adjust to changing circumstances with more noteworthy exactness and productivity. By tackling the force of computer-based intelligence, makers can open new degrees of efficiency, quality, and development in their operations.In expansion to progressions in mechanical technology equipment and programming, the reception of advanced advances like the Web of Things (IoT) and distributed computing is driving further development in assembling. These advancements empower robots to interface and speak with different machines, sensors, and frameworks in the creation climate, making interconnected biological systems known as brilliant plants. In savvy production lines, robots

can flawlessly trade information, coordinate undertakings, and answer continuous criticism, prompting more light-footed and responsive assembling processes.Moreover, mechanical technology in the industry isn't restricted to conventional assembling areas but on the other hand is venturing into new outskirts like added substance fabricating, otherwise called 3D printing. 3D printing robots can make complex calculations and custom parts with high accuracy and productivity, changing how items are planned, prototyped, and made. From aviation parts to clinical inserts, 3D printing robots offer makers uncommon adaptability and imagination in item advancement and production.As mechanical technology innovation proceeds to develop and develop, the limits between the physical and computerized universes are turning out to be progressively obscured, bringing about additional opportunities for advancement and coordinated effort. From independent robots and versatile robots for strategies and warehousing to mechanical frameworks for customized fabricating and on-request creation, the eventual fate of advanced mechanics in the industry holds boundless potential for changing how we configure, make, and convey products.In the end, advanced mechanics in industry are reshaping the scene of assembling and creation, empowering producers to accomplish new degrees of proficiency, adaptability, and development. By embracing the most recent headways in advanced mechanics innovation and utilizing the force of robotization, man-made reasoning, and computerized availability, makers can make dexterous and responsive creation frameworks that drive financial development, supportability, and seriousness in the worldwide commercial center. As we keep on investigating the potential outcomes of advanced mechanics in the industry, let us stay focused on outfitting innovation to support mankind, making a future where robots and people cooperate agreeably to fabricate a superior world for all. The combination of advanced mechanics in industry isn't just reshaping producing processes yet in addition setting out new open doors for monetary development and seriousness on a worldwide scale. By embracing advanced mechanics innovation, makers can smooth out creation, diminish costs, and further develop item quality, permitting them to stay light-footed and responsive in an undeniably aggressive commercial center. This, thus, can prompt an expanded piece of the pie, extended client bases, and more noteworthy productivity for organizations that embrace automation.Furthermore, advanced mechanics in the industry can drive development and business ventures by bringing hindrances down to sections and empowering little and medium-sized undertakings (SMEs) to contend with bigger partnerships. With the accessibility of reasonable and open automated frameworks, new companies and trailblazers can foster new items, investigate specialty advertisements, and disturb customary businesses with inventive arrangements. This democratization of advanced mechanics innovation cultivates a culture of development and imagination, prodding financial development and occupation creation in different areas of the economy.Moreover, the advantages of advanced mechanics in the industry reach out past monetary contemplations to envelop ecological maintainability and social obligation. By streamlining asset use, limiting waste, and diminishing energy utilization, mechanical technology-empowered assembling cycles can add economic and harmless to the ecosystem's future. Furthermore, via robotizing unsafe or genuinely requesting undertakings, robots assist with further developing working environment security and diminish word-related wounds and ailments, improving the prosperity and personal satisfaction of workers.As we plan, the capability of advanced mechanics in industry to drive positive change and change is boundless. From speeding up the speed of mechanical advancement to setting out new open doors for the monetary turn of events and social advancement, mechanical technology can shape the world in significant and significant ways. By embracing the most recent progressions in advanced mechanics innovation and cultivating coordinated efforts between industry, the scholarly community, and government, we can open the maximum capacity of mechanical technology to make a superior, more prosperous, and practical future for all. In the end, mechanical technology in industry addresses an extraordinary power that is reforming how products are fabricated, disseminated, and consumed. By saddling the force of computerization, man-made reasoning, and advanced networks, makers can make coordinated, effective, and responsive creation frameworks that drive monetary development, development, and supportability. As we keep on investigating the potential outcomes of mechanical technology in the industry, let us stay focused on tackling innovation to help mankind, making a future where robots and people cooperate amicably to construct a superior world for a long time into the future.

From Sequential construction systems to Shrewd Production lines

Moderate advancement frameworks: Critical stretches from times gone past Moderate improvement structures adjusted assembling during the 20th hundred years. Henry Part's presentation of the extraordinary improvement structure for capably making vehicles on an astoundingly key level affected capacity and cost-sensibility. By limiting complex endeavors to extra bona fide, dull advances, the Part's ever-evolving improvement structure thought about speedier creation and the game plan of the sensible Model vehicle. Progress of Get-together Computerization. Robotization (1800s to mid-1900s): Basic machines like pulleys and switches robotized intriguing work.

The reliably forming improvement structure changed into an indication of this stage, attracting monster expansion assembling and cost decline Robotization Levels of progress (1970s): Programmable reasoning controllers (PLCs) and PC numerical control (CNC) machines brought precision and flexibility. Producers could robotize more tangled processes. Amazing Get-together (Most recent thing): Capable plants coordinate model setting levels of headway like huge level mechanics, man-made thinking (modernized thinking), and the Catch of Things (IoT). These interconnected plans spread out autonomous creation conditions. Sharp assembling upgrades entire stock chains, from thing mean to move, using data appraisal and relentless taking notes. Benefits of Computerization in Party. Expanded Ability: Robotization speeds up creation, diminishing a doorway to incorporate for stock. Cost Reduction: Limiting disturbing work and goofs decreases costs. Regulated Quality: Robotization ensures solid quality by lessening unsteadiness. Redesigned Security: Less manual endeavors mean fewer risks. Marvelous Regulating plants versus Standard Mechanical creation systems Current working environments: Utilize interconnected plans and stuff to pass on driving data. Connect better choice creation for heads, controllers showed prepared experts and pioneers. Coordinate obvious level unquestionable level mechanics, man-made data, and IoT. Based on structure interconnectivity and data sharing. Need to reduce rejects, cut costs, and further assistance limit. Standard Mechanical creation structures: Integrate direct cycles where each expert performs unequivocal undertakings. This can actuate bottlenecks and deferrals. Come up short on flexibility and versatility of wonderful creation lines. End Sharp overseeing plants address the peak of get-together turn of events, using development to revive creation and supply chains. As we push ahead, the trustworthy coordination of physical and certain level universes will continue to shape the possible fate of get-togethers.

Chapter 4: Robots in Medical Care: Changing Medication and Patient

As of late, mechanical technology has arisen as a strong power in the field of medical care, reforming how operations are performed and how patient consideration is conveyed. From careful robots that help specialists with accuracy and adroitness to mechanical frameworks that give help and backing to patients, the combination of mechanical technology in medical services has prompted critical headways in therapy results, patient wellbeing, and by and large nature of care. In this section, we will investigate the effect of robots in medical care and the extraordinary job they play in molding the eventual fate of medicine. At the front line of mechanical technology in medical care are careful robots, which have changed the act of a medical procedure by offering exceptional degrees of accuracy, control, and perception. These mechanical frameworks are outfitted with cutting-edge imaging innovations, for example, top-quality cameras and 3D imaging, that furnish specialists with upgraded permeability and profundity insight during systems.

Also, automated arms with different levels of opportunity and aptitude empower specialists to perform complex moves with more prominent precision and adaptability than customary careful techniques. One of the most notable instances of carefully advanced mechanics is the da Vinci Careful Framework, which has been broadly taken on for negligibly obtrusive systems in claims to fame like urology, gynecology, and general a medical procedure. The da Vinci framework comprises of mechanical arms constrained by a specialist console, considering exact developments and sensitive tissue control with insignificant entry points. By limiting injury to encompassing tissues and organs, mechanical helped a medical procedure offer patients quicker recuperation times, diminished torment, and further developed corrective results contrasted with customary open surgery. In expansion to careful mechanical technology, robots are likewise assuming an undeniably significant part in clinical help and restoration. For instance, mechanical exoskeletons are being utilized to help patients with versatility impedances, like spinal line wounds or stroke, by offering fueled help to their lower appendages. These exoskeletons empower patients to stand, walk, and perform exercises of day-to-day living with more prominent freedom and certainty, prompting enhancements in actual capability and nature of life. Moreover, robots are being conveyed in telemedicine applications to give far-off counsel and observation to patients in underserved or far-off regions. Telepresence robots outfitted with cameras and screens permit medical care suppliers to connect with patients and lead assessments continuously, crossing over geological obstructions and growing admittance to medical care administrations. This is especially important in rustic networks or during crises when admittance to clinical consideration might be limited. Furthermore, robots are being utilized in an assortment of other medical care settings, including drug stores, research facilities, and restoration focuses, to mechanize routine undertakings and further develop productivity. Robotized medicine administering frameworks guarantee precise dosing and diminish the gamble of prescription blunders, while automated phlebotomy gadgets smooth out blood assortment systems and limit inconvenience for patients. Furthermore, robots are being utilized in non-intrusive treatment and recovery to give customized activities and treatment meetings custom-made to individual patient needs.However, as advanced mechanics innovation keeps on propelling, it additionally raises moral, administrative, and cultural ramifications that should be tended to. Worries about understanding well-being, security, and risk require cautious thought and oversight to guarantee that robots are conveyed capably and morally. Furthermore, endeavors to address variations in admittance to automated innovation and medical services administrations are fundamental to guarantee that all patients benefit from the capability of advanced mechanics to work on clinical results and the nature of life.In the end, robots are changing the scene of medical services, offering new open doors for working on operations, patient consideration, and in general well-being results. From careful robots that empower negligibly intrusive strategies to automated exoskeletons that help with versatility and restoration, the mix of mechanical technology in medical services is opening up new wildernesses for advancement and revelation. As we keep on investigating the capability of robots in medical care, let us stay directed by our obligation to propel human prosperity and make a future where innovation serves the requirements of patients and medical services suppliers alike. Moreover, as the field of mechanical technology in medical services keeps on developing, new developments and applications are arising that guarantee to additionally change the act of medication and patient consideration. One such area of advancement is the utilization of computerized reasoning (artificial intelligence) and AI calculations to upgrade the abilities of automated frameworks. Overwhelmingly of clinical information, artificial intelligence-controlled robots can help clinicians in diagnosing sicknesses, arranging therapy systems, and anticipating patient results with more prominent precision and efficiency. Additionally, propels in sensor advancements and wearable gadgets are empowering the improvement of customized medication and distant patient observing arrangements. For instance, robots furnished with biosensors and physiological checking gadgets can follow imperative signs, identify early admonition indications of ailments, and give opportune medications or alarms to patients and medical care suppliers. This continuous checking and input empower proactive administration of persistent infections and diminishes the requirement for regular emergency clinic visits, prompting work on tolerant results and cost reserve funds for medical care systems.Furthermore, advanced mechanics are upsetting the field of clinical imaging and diagnostics, considering more exact and productive recognition of sicknesses and irregularities. Automated imaging frameworks, for example, X-ray-directed robots and mechanical ultrasound scanners, empower exact focus on and perception of physical designs, upgrading analytic precision and diminishing the requirement for intrusive methodology. Furthermore, mechanical biopsy gadgets empower clinicians to get tissue tests with more noteworthy accuracy and negligible gamble to patients, prompting more precise determinations and treatment planning.Moreover, mechanical technology is assuming a pivotal part in tending to basic medical care difficulties, like the Coronavirus pandemic, by empowering fast turn of events and organization of demonstrative tests, therapeutics, and immunizations. Robots are being utilized in labs to mechanize high-throughput testing processes, speeding up the discovery of viral diseases and working with contact following endeavors. Also, robots are being conveyed in clinics to sanitize surfaces, convey drugs, and help with patient consideration, lessening the gamble of transmission and easing the weight on medical care workers. However, as robots become progressively coordinated in medical care settings, it is fundamental to address concerns connected with patient protection, information security, and moral contemplations. Shields should be set up to guarantee that patient data is safeguarded and that robots are utilized mindfully and morally as per laid-out clinical rules and guidelines. Furthermore, endeavors to address aberrations in admittance to mechanical innovation and medical care administrations are urgent to guarantee fair medical services conveyance and further develop well-being results for all patients. In the end, mechanical technology is ready to change the act of medication and patient consideration in significant and effective ways. From careful robots that empower insignificantly obtrusive methods to computer-based intelligence-controlled symptomatic frameworks and far-off understanding checking arrangements, the joining of mechanical technology into medical services holds colossal commitment for working on clinical results, lessening medical services costs, and improving the personal satisfaction of patients. As we keep on investigating the capability of robots in medical care, let us stay focused on tackling innovation to help humankind, making a future where everybody approaches top caliber, merciful, and customized medical services administrations. Besides, as robots become dependably arranged into clinical advantages frameworks, it is vital to focus on the interdisciplinary joint effort and decoration commitment to guarantee that electronic drives address the issues and notions of patients, clinical thought suppliers, and different accomplices. By making the relationship between engineers, clinicians, trained professionals, policymakers, and patients, we can co-make imaginative arrangements that address the puzzling difficulties and entrances in clinical advantage transport. Likewise, endeavors to push preparing and organizing in mechanical new development and clinical thought are epic for setting up the striking new season of clinical thought arranged specialists and technologists to set up the best furthest extents of automated progress. By giving doorways for a dynamic experience, interdisciplinary joint effort, and reliable learning, we can furnish clinical advantages experts with the information and endpoints they need to arrange mechanical movement into their clinical practice and further help patient idea results. Moreover, as we plan, proceeding to put resources into creative endeavors to move the best in class in mechanical movement and clinical ideas is crucial. By supporting interdisciplinary assessment projects, development move drives, and public-private affiliations, we can speed up the speed of progress and bring incredibly electronic pushes from the lab to the workplace. This merges growing new motorized stages, assessments, and sensors that address acquired clinical necessities and empower re-endeavored, patient-focused care. At long last, mechanical improvement is ready to change the showing of medication and patient contemplations, offering new paths for directing clinical results, redesigning patient encounters, and diminishing clinical advantages costs. By embracing the limitation of mechanical improvement in clinical thought and working consistently across disciplines and regions, we can make a future where everybody approaches undeniable grade, smart, and changed clinical advantages affiliations. As we keep on looking at the regular results of robots in clinical thought, let us stay directed by our obligation to push individuals to succeed and make a future where improvement serves the necessities of patients and clinical advantages suppliers the same.

Advancements in Careful Mechanical Technology and Clinical Help

Robot-Helped Medical Procedures: Robot-helped medical procedures have advanced since they originated in the last part of the 1960s. Present-day automated careful frameworks come furnished with profoundly adroit arms and scaled-down instruments. These frameworks decrease quakes, empower fragile moves, and upgrade careful exactness. The combination of imaging and representation advances further develops accuracy. Haptic Criticism Framework: Careful robots currently integrate a haptic input framework.

➢ This permits specialists to survey tissue consistency during strategies without actual contact, forestalling wounds because of unreasonable power application. Teleoperation: Specialists can defeat topographical limits by utilizing teleoperation. This innovation empowers particular medical care conveyance from a distance. Computerized reasoning (computer-based intelligence) and AI (ML): computer-based intelligence and ML assume a critical part in careful direction. They upgrade the acknowledgment of perplexing physical designs, prompting improved results for patients. Quicker Recuperation and Less Confusions: This multitude of headways adds to quicker persistent recuperation and less post-careful complexities. Be that as it may, there are difficulties to survive: Cost: Mechanical frameworks are costly to secure and keep up with.

➤ Size: The size of mechanical frameworks can impede certain settings. Specialist Preparing: Legitimate preparation is fundamental for the successful utilization of careful robots. Regardless of these difficulties, the fate of mechanical medical procedures looks encouraging. Advancements, for example, man-made intelligence-driven mechanization, nanorobots, tiny cut medical procedures, semi-robotized telerobotic frameworks, and the effect of the 5G network on far-off medical procedures keep on driving advancement in medical services. Organizations like Natural Careful, Johnson and Johnson, Medtronic, and Olympus are driving pioneers in this field.

Chapter 5: The Job of Robots in Investigation: Propelling Space and Maritime Disclosures

Robots have long perceived a fundamental part in making discernment we could loosen up the universe and open the secrets of dull locale, both on the planet no inquiry. From electronic drifters crossing the Martian surface to free chop-down vehicles planning the profundities of the sea profundities, the mechanical examination is increasing the requirements of human information and reshaping our point of view on the universe.

In this part, we will look at the changed control of robots in assessment and the crucial openings they participate in space and the oceans.At the extreme forefront of mechanical assessment is the field of room mechanical development, which wraps up endless motorized missions and advances that are expected to examine divine bodies and check the universe out. Computerized strays, for example, NASA's Mars vagabonds Soul, Opportunity, and Premium, have changed how we could unravel the Red Planet by inspecting its surface, driving reasonable assessments, and get-together geographical models. These wanderers are furnished with a set-up of instruments, including cameras, spectrometers, and drills, that draw in them to look at the Martian scene and excursion for indications of past or present life.Furthermore, mechanical rockets, for example, NASA's Explorer tests and the Mars strays, have wandered past our planetary get-together, giving major information and pieces of data into the external levels of the universe.

These space mechanical social occasions are furnished with sensors and instruments that award them to zero in on distant planets, moons, and magnificent unconventionalities, uncovering a discernment of the outline and improvement of our planetary party and the more fundamental universe. Furthermore, mechanical telescopes and observatories, for example, the Hubble Space Telescope and the James Webb Space Telescope, keep on changing insight we could unravel the universe by finding shocking pictures and assembling information from far-off structures and self-important phenomena.In expansion to space assessment, robots are correspondingly expecting a significant part in the ocean evaluation, connecting with experts to study and guide the colossal and pardoned profundities of the sea, as a rule, base. Free cut-down vehicles (AUVs) and remotely worked vehicles (ROVs) outfitted with cameras, sonar, and different sensors are ready for dropping to profundities of thousands of meters, gathering fundamental standard information and symbolism of chopped-down scenes and ordinary plans. These robots draw in experts to zero in on far-off ocean watery vents, coral reefs, and marine life, giving key pieces of data into the interconnectedness of Earth's seas and the effect of human exercises on marine ecosystems.Moreover, robots are being sent in insane conditions, for example, the polar region and distant ocean channels to work with sensible appraisal and screen typical changes. Ice-entering robots, like NASA's Icebreaker, are utilized to zero in on the bits of polar ice sheets and screen changes in ocean level and environment. In addition, far away ocean ROVs outfitted with controller's arms and seeing instruments draw in experts to assemble essentials of headway, rock, and marine life from the sea profundities, adding to how we could unravel Earth's space history and biodiversity.Furthermore, certain level mechanics development is drawing in the improvement of creative reactions for investigating and colonizing other stunning bodies, like the Moon and Mars. Automated landers and conditions equipped with a presence of really important affiliations and asset use pushes are being made to help human assessment missions to these far-off universes. Moreover, free robots and vagabonds are being considered for use in the development of lunar and Martian ordinary typical parts, as well as concerning prospecting and mining essential assets like water and minerals.However, as we experience further into space and investigate the profundities of the sea, it is critical to think about the moral, normal, and legitimate delayed consequences of robotized assessment. Attempts to save and watch mind-blowing bodies and ocean life customary plans against debasing and flabbergasting require a wary procedure and coordination among by and large relates. Moreover, worries about space waste and pollution should be addressed to guarantee the realness of room examination exercises and cut off the bet of contact with obliging rockets and satellites.In certification, robots are anticipating an essential part of driving seeing so we can translate the universe and foster the woodlands of human assessment. From analyzing faraway planets and dynamite bodies to organizing the profundities of the sea profundities, the mechanical examination is related to fundamental openings and reshaping how we can translate the universe. As we keep on growing the limitations of mechanical examination, let us stay worked with by our most memorable rate, imaginative frontal cortex, and obligation to research the weak and loosely held bits of information of the universe.Moreover, as progress keeps on impacting, the imperatives of robotized travelers ought to turn out to be further, stunning in areas of strength for extra and divulgences in both space and ocean assessment. For example, future space missions could integrate the arrangement of enormous measures of restricted scope robots to evaluate planetary surfaces impressively more rapidly, amass tests, and direct appraisals straightforwardly. These robots could share kindly, passing and figuring out their activities to accomplish discerning targets with more capacity than individual missions.Similarly, in ocean assessment, kinds of progress in top-tier mechanics progress are opening up additional passageways for zeroing in on great conditions, for example, watery vents, far away ocean channels, and ice-covered seas. Scaled-back AUVs outfitted with cutting-edge sensors and dismantling devices could be given in colossal numbers to plan and analyze these remote and testing to show up at districts, revealing information about the biodiversity, geography, and standard states of the huge ocean. Furthermore, certain level mechanics are working with overall assistance and participation in assessment attempts, with space affiliations, research establishments, and restrictive affiliations consolidating attempts to pool assets and fortitude to supervise complex solid inconveniences. For instance, the General Space Station (ISS) finishes as a stage for arranging preparations and testing overhauls in a microgravity climate, with space explorers and mechanical plans taking part to incite how we could translate human thriving, materials science, and space assessment technologies. Similarly, consistent drives, for example, the Sea Assessment Trust's Nautilus Assessment Program joins arranged specialists, fashioners, and teachers from around the world to research and remain a lot early dull district of the sea profundities. By utilizing robotized progress, for example, ROVs and AUVs, these endeavors are uncovering new species, land techniques, and typical structures, updating how we could loosen up the marine climate and its significance to life on Earth. Moreover, as mechanical assessment limits keep on improving, there is making pay to utilize robots to look for indications of extraterrestrial life and endurable conditions on different planets and moons. Missions to cold moons, for example,

Europa and Enceladus, which could make sense of subsurface seas under their frozen surfaces, could harden sending mechanical tests to analyze these far-off universes and outing for check of microbial life or conditions reliable for life as we know it. However, as we leave on areas of strength for these missions, it is fundamental to address the moral, guaranteed, and social repercussions of mechanized evaluation. Demands concerning planetary security, typical effect, and the reasonable scattering of assets should be carefully considered to guarantee that assessment rehearses are made continually and as per general procedures and frameworks. In addition, endeavors to draw in people all around and energize discussion of the advantages and dangers of mechanical assessment are fundamental for building sponsorship and understanding for future evaluation endeavors. In the decision, robots are getting through an uncommon part in enacting care we could unravel the universe and grow the instigated areas of human examination. From investigating faraway planets and exceptional bodies to arranging the profundities of the sea profundities, mechanical pilgrims are opening new openings and reshaping how we can unravel the universe. As we keep on extending the limitations of robotized assessment, let us stay shaped by our five-star, creative mind, and obligation to explore the dull and prod people in the future to endeavor the unfathomable.

From Mars Wanderers to Remote Ocean Voyagers

Right when we hear "stray," our minds continually leap to pictures of Mars evaluation, where mechanical vagabonds like Steady Quality and Premium explore the Red Planet's surface, dismantling its geology for signs of past reasonableness. Regardless, Earth moreover shows its strays, and they research a substitute wild: the immense ocean. One such memorable vagabond is the Benthic Wanderer II, made by experts at the Monterey Delta Aquarium Assessment Connection (MBARI). By no means like its Martian partners, Benthic Vagabond II works 4,000 meters under the ocean surface, on a new critical plain, beating the stunning pile of 6,000 pounds for each square inch of pressure. We ought to hop into the enchanting universe of distant sea evaluation and study this confusing stray. Benthic Drifter II: Researching the Critical Sea Base Carbon Cycling Assessment: Benthic Stray II's key mission is to accumulate data related to carbon cycling. It searches for answers to questions like What carbon sources show up at the far-off sea profundities? Does that carbon return to the environment as carbon dioxide (potentially adding to an overall temperature change), or does it remain safely sequestered in ocean improvement? By studying oxygen use by base-standing animals and microorganisms after some time, the wanderer helps scientists with understanding how carbon moves from the surface to the sea base. Testing Environment: The far-off sea environment where Benthic Wanderer II works is past silly: Critical plain: An ignored, tumultuous sea base at an importance of 4,000 meters. Cold temperatures and high strain: The wanderer progresses forward through freezing conditions and enormous pressure.

> Darkness: Sunlight doesn't enter these profundities, so the vagabond relies on counterfeit lighting. Free Evaluation: Benthic Vagabond II works uninhibitedly, investigating the sea base, getting photographs, and assembling data. Its camera gets shuddering encounters with huge fishes, for instance, looking through rattails (Coryphaenoides sp.). Contemplations for Ecological Change: Understanding carbon cycling in the distant sea has more significant

repercussions for ordinary change. Expecting carbon dioxide to be liberated from the sea base, could add to all-around warming. Then again, sequestering carbon in ocean development mitigates ordinary impacts. Organizing Burdens: Making a stray for the distant sea incorporates astounding orchestrating snags: Liberal materials: The drifter ought to move beyond crazy pressure and heartbreaking saltwater. Clear course: Scene relative course, similar to what worked with the Mars stray, helps Benthic Wanderer II with examining truly. In summation, while Mars drifters break down distant planets, Benthic Wanderer II jumps into the mysteries of our colossal oceans. Its data adds to how we can unwind carbon parts and enlightens our method for managing and administering normal stewardship.

Chapter 6: Advanced Mechanics and Instruction: Forming the Fate of Learning

Of late, mechanical innovation has emerged as a necessary resource for evolving and preparing, offering students of all ages the opportunity to participate in elaborate learning experiences that support creative minds, conclusive thinking, and decisive reasoning skills. From grade schools to universities, high-level mechanics programs are energizing students to explore science, development, planning, and math (STEM) disciplines in imaginative and associating ways.

In this part, we will research the occupation of mechanical innovation in preparing and its impact on trimming the destiny of learning.At the center of mechanical innovation, tutoring is the perspective of progressing by doing, where students successfully take part in arranging, building, and programming robots to handle genuine hardships. By working helpfully in gatherings, students procure huge capacities like correspondence, joint exertion, and erranding the board, which are major results in the 21st-century workforce. Besides, mechanical innovation projects support inventiveness and improvement, as students are encouraged to investigate various roads regarding different plans and deals to achieve their goals. One of the most notable stages for mechanical innovation preparation is LEGO Mindstorms, which outfits students with an adaptable and straightforward stage for building and programming robots using LEGO blocks and sensors. LEGO Mindstorms packs consolidate programmable blocks, motors, sensors, and programming gadgets that engage students to plan and create robots that can play out an enormous number of endeavors, from investigating obstruction courses to orchestrating things or playing. These units are used in homerooms all around the planet to show students the fundamentals of mechanical innovation and programming in a silliness and shrewd way. Furthermore, mechanical innovation contentions like FIRST Mechanical Innovation and VEX Progressed Mechanics offer students the possible opportunity to apply their capacities and data in a relentless setting, where they design, develop, and program robots to fight in a movement of challenges. These challenges outfit students with involved understanding as well as develop collaboration, sportsmanship, and a sensation of improvement as gatherings crew up to deal with incredible issues and achieve shared goals. Besides, mechanical innovation contentions offer students receptiveness to authentic planning practices and industry mentors, giving significant encounters into potential business pathways in STEM fields.Moreover, mechanical innovation preparation isn't confined to customary review corridor settings however then again is being facilitated in relaxed learning conditions, for instance, after-school programs, day camps, and maker spaces. These relaxed learning open entryways license students to examine progressed mechanics at their speed and seek after their tendencies in STEM subjects past the homeroom. Besides, high-level mechanics clubs and affiliations outfit students with a sensation of neighborhood having a spot, where they can collaborate with peers who share near interests and passions.Furthermore, high-level mechanics expect a basic part in propelling assortment and thought in STEM preparation by giving entryways to underrepresented get-togethers, including women and minorities, to pursue part in dynamic open doors for development and research calling pathways in development and planning. Drives, for instance, Young Women Who Code and Ethnic minorities CODE are endeavoring to connect with young women and young women to seek after callings in STEM fields through cutting-edge mechanics and coding programs that accentuate the creative mind, facilitated exertion, and authority development. However, as cutting-edge mechanics tutoring continues to develop, it is fundamental to address challenges like access, worth, and teacher planning to ensure that all students have the expected opportunity to benefit from cutting-edge mechanics preparation. Attempts to assemble permission to cutting-edge mechanics resources and activities in underserved networks, give capable improvement significant entryways to educators, and advance far-reaching showing practices are basic for closing the STEM direction opening and drawing in the exceptional time of trailblazers and issue solvers. In the end, mechanical innovation is changing tutoring by offering students dynamic chances for development that empower creative minds, definitive thinking, and composed exertion. From LEGO Mindstorms packs in grade schools to mechanical innovation challenges in optional schools and universities, high-level mechanics preparation is stirring students to explore STEM subjects in up to this point unfathomable ways. As we continue to furnish the power of cutting-edge mechanics in preparing, let us remain fixed on laying out extensive learning conditions that connect all students to succeed and prosper in the 21st century. Moreover, as advancement continues to create, the open doors for cutting-edge mechanics in tutoring are developing, offering new entryways for striking and tweaked open doors for development. Virtual and extended reality (VR/AR) progressions, for example, are being facilitated into mechanical innovation preparing to lay out virtual circumstances where students can design, build, and test robots in reenacted settings. These virtual experiences engage students to examine complex thoughts and circumstances in a safeguarded and natural manner, updating their understanding and support of STEM principles. Additionally, mechanical innovation is being used to help interdisciplinary progress across many parts of information, from craftsmanship and music to history and composing. For example, high-level mechanics that solidify parts of describing, inventiveness, and design challenge students to think in a general sense and imaginatively as they revive their contemplations through cutting-edge mechanics. By planning progressed mechanics into grouped curricular settings, educators can attract students in huge and significant open doors for development that beat any boundary between speculation and practice.Furthermore, high-level mechanics empower overall participation and social exchange by partner students from different countries and establishments through shared progressed mechanics adventures and contentions. Projects, for instance, the Vital Overall Test and RoboCup Junior join gatherings of students from around the world to collaborate on mechanical innovation hardships and display their gifts on an overall stage. These worldwide composed endeavors advance diverse cognizance and partnership as well as allow students critical opportunities to encourage participation, correspondence, and organization capacities in a multicultural context.Moreover, mechanical innovation tutoring connects students to become issue solvers and improve their organizations by applying their understanding and capacities to determine genuine issues and troubles. For example, mechanical innovation projects focused in on regular safeguarding, failure response, and clinical consideration engage students to include progressed mechanics development for social extraordinary and have a valuable result on their organizations. By partaking in assist learning projects, students with fostering compassion, sympathy, and a sensation of social commitment, getting them positioned to become moral and associated with occupants in an irrefutably interconnected world.However, as mechanical innovation continues to be created, it is fundamental to address stresses over the moral, social, and biological consequences of cutting-edge mechanics development. Discussions about the ethical use of mechanical innovation, including issues like security, freedom, and inclination, ought to be facilitated into cutting-edge mechanics instructive arrangement to ensure that students make a nuanced cognizance of the ethical examinations drawn in with arranging and sending computerized structures. Additionally, attempts to propel sensibility and proficient improvement in mechanical innovation tutoring are key for ensuring that students are prepared to address the puzzling challenges and chances of the future.In the end, high-level mechanics is changing preparation by offering students attractive and striking chances for development that develop creative minds, unequivocal thinking, and composed exertion. From LEGO Mindstorms units in grade schools to worldwide mechanical innovation challenges in optional schools and universities, mechanical innovation preparation is moving students to examine STEM subjects in up until now unfathomable ways. As we continue to saddle the power of mechanical innovation in preparing, let us keep fixed on laying out complete learning conditions that connect all students to turn out to be well-established understudies and pioneers who can prosper in the 21st hundred years to say the very least.

Coordinating Mechanical technology into STEM Educational program

Sorting out mechanical improvement into STEM (Science, Progression, Collecting, and Finding out) arranging is head for assembling understudies with the endpoints they need for the general world. We should look at how improvement can empower STEM learning: Online Standard Learning Conditions: These stages award understudies to be drawn to satisfaction. They can participate in reenactments, tests, and obliging exercises related to mechanical appraisals. Online instruments can give quick data and change as per individual driving basics. Redirection: Augmentations are central assets for showing mechanical rules. Understudies can endeavor different things in various conditions, notice results, and gain sensible experiences. For instance, replicating mechanical plans or figuring out virtual models can stay aware of their comprehension. Expanded Reality (AR): AR overlays advanced data onto this strong reality. In a mechanical setting, AR can assist understudies with imagining complex plans, like motors or stuff structures. Envision understudies wearing AR glasses and seeing commonplace 3D models of mechanical parts during a model. PC-made Reality (VR): VR hacks down understudies in a PC-conveyed climate. For mechanical heading, VR can reenact making plant floors, moderate improvement plans, or much space. Understudies can see gear, research issues, and practice support tries in a got, controlled climate. Electronic Gaming: Gamification can make learning mechanical evaluations indisputable. Illuminating games can start understudies to sink into coordinating issues, gathering structures, or advancing mechanical frameworks. By organizing game mechanics, instructors can remain mindful of realness and inspiration.

Chapter 7: Independent Vehicles: Heading Toward a Driverless Future

As of late, independent vehicles have arisen as a groundbreaking innovation with the possibility to reform how we travel, drive, and transport products. From self-driving vehicles and trucks to independent robots and conveyance robots, the ascent of independent vehicles is reshaping the fate of transportation and versatility. In this part, we will investigate the turn of events, difficulties, and ramifications of independent vehicles as we head toward a driverless future.

At the bleeding edge of independent vehicle innovation are self-driving vehicles, which utilize a blend of sensors, cameras, radar, and man-made consciousness calculations to explore streets and traffic without human mediation. Organizations like Tesla, Waymo, and Journey are driving the way in creating and testing independent driving frameworks that guarantee to make streets more secure, decrease gridlock, and increment portability for individuals of any age and capacity. These self-driving vehicles can change metropolitan transportation, empowering on-request versatility benefits and shared independent vehicle armadas that supplement public travel and lessen dependence on confidential vehicle ownership.Furthermore, independent vehicles are ready to alter the coordinated factors and transportation industry by empowering completely independent trucks and conveyance vehicles that can work day in and day out without the requirement for human drivers. Organizations, for example, Leave, TuSimple, and Amazon are creating independent shipping arrangements that guarantee increment proficiency, lessen costs, and further develop well-being in long-stretch cargo transportation. Via robotizing routine undertakings, for example, driving and routing, independent trucks can change production network planned operations and upset how merchandise is shipped and conveyed the nation over and around the world.Moreover, independent vehicles are growing past conventional street transportation to incorporate automated aeronautical vehicles (UAVs) and drones that can independently explore airspace and convey labor and products to remote or unavailable regions. Organizations, for example, Amazon Prime Air and Google's Wing are creating independent robot conveyance frameworks that guarantee to reform last-mile planned operations and empower quicker, more effective conveyance of bundles, clinical supplies, and crisis reaction administrations. These independent robots can change ventures like web-based businesses, medical services, and calamity help by giving quick, on-request conveyance to clients and networks in need.However, as independent vehicles become progressively coordinated into our transportation frameworks, they additionally bring up significant issues and difficulties connected with security, guidelines, and morals. Worries about the unwavering quality and well-being of independent driving frameworks, the potential for mishaps and impacts, and the moral ramifications of programming choices should be painstakingly addressed to guarantee that independent vehicles are sent dependably and morally. Also, endeavors to lay out clear administrative structures and guidelines for independent vehicle testing and organization are fundamental for guaranteeing public trust and trust in this rising technology.Furthermore, as independent vehicles become more pervasive on our streets and in our skies, they can reshape metropolitan scenes and change how we plan and plan urban communities. Independent vehicles might prompt changes in land use, leaving framework, and transportation organizations, as urban areas adjust to oblige new methods of versatility and diminish dependence on confidential vehicle proprietorship. Furthermore, independent vehicles can further develop admittance to transportation for underserved networks, decrease ozone-depleting substance emanations, and set out new open doors for financial turn of events and social equity.In the end, independent vehicles are driving us toward a future where transportation is more secure, more proficient, and more open for all. From self-driving vehicles and trucks to independent robots and conveyance robots, the ascent of independent vehicles is reshaping how we move merchandise and individuals, offering new open doors for development and disturbance in the transportation business. As we keep on exploring the street toward a driverless future, let us stay aware of the valuable open doors and difficulties that independent vehicles present, and work together to guarantee that this extraordinary innovation benefits society as a whole.Moreover, as independent vehicle innovation keeps on propelling, there is developing interest in investigating its expected applications in different areas past transportation, including farming, development, and public security. Independent robots and robots, outfitted with sensors and artificial intelligence calculations, are being utilized to screen crops, assess foundations, and answer crises in remote or dangerous conditions. These independent frameworks offer new open doors for expanding efficiency, lessening costs, and further developing security in an extensive variety of industries.Additionally, independent vehicles can change how we ponder portability and availability for individuals with handicaps and versatility challenges. Self-driving vehicles and independent transports furnished with wheelchair-available highlights and assistive advances offer additional opportunities for autonomous travel and local area reconciliation for individuals with inabilities. By giving on-request, house-to-house transportation administrations, independent vehicles can upgrade the personal satisfaction and social consideration for individuals with portability impairments.Furthermore, as independent vehicles become more common on our streets and in our urban communities, they are creating huge measures of information that can be utilized to further develop transportation frameworks and metropolitan preparation.

By examining information gathered from sensors, cameras, and different sources, transportation organizers and policymakers can acquire experiences in traffic designs, blockage areas of interest, and travel conduct, empowering them to arrive at informed conclusions about framework ventures and transportation approaches. Furthermore, independent vehicles can speak with one another and with shrewd foundation frameworks to improve traffic stream, lessen mishaps, and upgrade by and large transportation efficiency.However, likewise, with any problematic innovation, the far and wide reception of independent vehicles additionally presents difficulties and potential dangers that should be tended to. Worries about network safety, protection, and information assurance should be addressed to guarantee the trustworthiness and security of independent vehicle frameworks and the information they create. Moreover, the progress to independent vehicles might have suggestions for business and work markets, especially for laborers in enterprises, for example, transportation and coordinated operations who might be uprooted via automation.Moreover, moral contemplations connected with dynamic calculations and moral problems should be painstakingly considered to guarantee that independent vehicles focus on human security and prosperity in all circumstances. Inquiries concerning risk and responsibility in case of mishaps or disappointments of independent vehicle frameworks likewise should be addressed to guarantee that fitting legitimate systems are set up to safeguard the privileges and interests of all gatherings involved. In the end, independent vehicles are driving us toward a future where transportation is more secure, more productive, and more open for all. From self-driving vehicles and trucks to independent robots and conveyance robots, the ascent of independent vehicles is reshaping how we move merchandise and individuals, offering new open doors for advancement and disturbance in the transportation business. As we keep on exploring the street toward a driverless future, let us stay aware of the potential open doors and difficulties that independent vehicles present, and work together to guarantee that this extraordinary innovation benefits society in general.

Navigating the Roads with AI-Powered Vehicles

The headway of Man-made consciousness (man-made intelligence) in the advancement of independent vehicles has changed how we imagine transportation. How about we investigate how artificial intelligence is forming the eventual fate of self-driving vehicles and making streets more secure and more productive? Human-Like Thinking for Independent Route: MIT specialists have made a framework that empowers driverless vehicles to explore new, complex conditions utilizing just basic guides and visual information.

Human drivers depend on perception and basic apparatuses to explore new streets. They match what they see around them to GPS data. Interestingly, driverless vehicles battle with this essential thinking. They should initially plan and examine new streets, which is tedious. The MIT framework "masters" guiding examples from human drivers as they explore a little region. It utilizes a camcorder takes care of and a straightforward GPS-like guide. When prepared, the framework has some control over a driverless vehicle along an arranged course in a shiny new region by mimicking the human driver. It likewise identifies confounds between its guide and street highlights, permitting it to address its course. Artificial intelligence Applications in Independent Vehicles: Simulated intelligence assumes a critical part in different parts of independent vehicles: Discernment: computer-based intelligence calculations decipher sensor information from cameras, lidar, radar, and different sensors to figure out the climate. Direction: simulated intelligence assists vehicles with pursuing split-subsequent options given sensor inputs, traffic conditions, and security contemplations. Sensor Combination: computer-based intelligence consolidates information from various sensors to make a far-reaching perspective on environmental factors. Planning and Limitation: simulated intelligence helps with making and refreshing guides, as well as deciding the vehicle's exact area. The objective is to accomplish a hearty independent route in new conditions. For instance, a framework prepared to drive in a metropolitan setting ought to easily explore lush regions it has never seen Security and Solace: Artificial intelligence calculations anticipate the activities of other street clients, guaranteeing safe collaborations. Self-driving vehicles constantly gain from new situations, adjusting to changing street conditions. By depending on man-made intelligence, independent vehicles upgrade security and give an agreeable travel insight to travelers.

Chapter 8: Advanced Mechanics and Farming: Developing Proficiency and Supportability

Lately, mechanical technology has arisen as a critical driver of development in horticulture, offering ranchers new instruments and advances to further develop efficiency, diminish work costs, and limit natural effects. From independent work vehicles and robots to mechanical collectors and weeders, the joining of mechanical technology into agribusiness is changing how yields are planted, tended, and reaped. In this section, we will investigate the job of mechanical technology in agribusiness and develop proficiency and supportability in food production potential. At the cutting edge of mechanical technology in horticulture are independent vehicles and robots that empower accurate cultivating methods, for example, factor rate cultivating, designated pesticide application, and yield observing. Independent work vehicles outfitted with GPS and sensors can explore fields with accuracy, sowing seeds and applying manures or pesticides with ideal precision and productivity. Also, drones furnished with cameras and sensors can gather high-goal symbolism and information on harvests, soil conditions, and field changeability, empowering ranchers to arrive at informed conclusions about the water system, preparation, and nuisance of the executives. Besides, advanced mechanics are reforming crop collecting and taking care of cycles, empowering quicker, more proficient reaping with diminished work necessities. Automated gatherers outfitted with vision frameworks and mechanical arms can specifically collect ready products of the soil with accuracy, limiting waste and expanding yield. Also, mechanical frameworks for arranging, evaluating, and pressing yields empower ranchers to process and bundle gathered produce rapidly and effectively, decreasing post-reap misfortunes and further developing item quality and period of usability. Also, mechanical technology is being utilized to address work deficiencies and rising work costs in horticulture via robotizing dull and truly requesting undertakings, for example, weeding, pruning, and diminishing. Mechanical weeders outfitted with cameras and man-made intelligence calculations can distinguish and eliminate weeds with accuracy, diminishing the requirement for compound herbicides and physical work. Essentially, mechanical pruning frameworks can manage plants and trees with accuracy, advancing natural product creation and decreasing work costs for cultivators. Also, advanced mechanics innovation is empowering the improvement of indoor cultivating frameworks like vertical ranches and tank-farming nurseries, where harvests are filled in controlled conditions under fake lighting and environment control frameworks. Independent robots and transport frameworks are utilized to move and oversee plants all through the developing system, from seedling engendering to gathering and bundling.

These indoor cultivating frameworks offer benefits, for example, all-year creation, higher harvest yields, and decreased water and pesticide utilization contrasted with customary open-air cultivating methods.However, as advanced mechanics innovation keeps on propelling, it additionally brings up significant issues and difficulties connected with reception, guidelines, and cultural ramifications. Worries about the expense and openness of advanced mechanics innovation for limited scope and family ranchers, as well as the potential for work removal in provincial networks, should be painstakingly considered to guarantee that the advantages of advanced mechanics in farming are impartially disseminated. Furthermore, endeavors to address moral and natural contemplations, like the utilization of pesticides and hereditary designing related to advanced mechanics, are fundamental for advancing supportable and dependable cultivating practices.In the end, advanced mechanics is changing horticulture by offering ranchers new devices and innovations to further develop effectiveness, efficiency, and manageability in food creation. From independent vehicles and robots for accuracy cultivating to automated gatherers and indoor cultivating frameworks, the combination of advanced mechanics in horticulture is changing how yields are developed, reaped, and made due. As we keep on saddling the force of advanced mechanics in horticulture, let us stay focused on advancing comprehensive and maintainable cultivating rehearses that benefit ranchers, buyers, and the climate alike.Moreover, as advanced mechanics innovation keeps on developing, there is developing interest in investigating its possible applications in addressing worldwide food security challenges and guaranteeing admittance to nutritious and reasonable nourishment for all. Mechanical technology-empowered arrangements like computerized vertical ranches, aquaculture frameworks, and aquaponic frameworks offer open doors for all-year food creation in metropolitan and peri-metropolitan regions, decreasing dependence on customary farming and expanding neighborhood food versatility. Moreover, advanced mechanics innovation can assume an urgent part in upgrading rural efficiency and flexibility notwithstanding environmental change, by empowering ranchers to adjust to changing natural circumstances and relieve the effects of outrageous climate events.Furthermore, mechanical technology is working with information-driven dynamics in horticulture by empowering ranchers to gather and dissect tremendous measures of information from sensors, drones, and different sources to enhance ranch the executives rehearses and further develop crop yields. By utilizing AI calculations and prescient investigation, ranchers can acquire experiences in crop wellbeing, soil richness, and weather conditions, empowering them to settle on informed conclusions about planting, water systems, and the nuisance of the executives. Furthermore, advanced mechanics innovation can empower ranchers to execute accurate agribusiness strategies, for example, site-explicit harvest the executives and variable rate application, enhancing asset use and limiting ecological impact. Additionally, mechanical technology is driving advancement in farming innovative work, empowering researchers and scientists to foster new harvest assortments, rearing procedures, and agronomic practices to further develop crop strength, nourishing quality, and yield. Advanced mechanics empowered phenotyping stages, for instance, empower scientists to quickly screen and assess a large number of plant characteristics, assisting with speeding up the rearing of harvests with further developed dry season resilience, sickness opposition, and healthful substance. Additionally, automated frameworks for plant reproducing and hereditary designing propositions open doors for exact and designated control of plant genomes to improve wanted qualities and attributes. Also, advanced mechanics innovation is cultivating cooperation and information trade among ranchers, specialists, and industry partners, through drives, for example, open-source advanced mechanics stages, producer spaces, and cooperative examination organizations. By sharing assets, aptitude, and best practices, partners can speed up the turn of events and reception of advanced mechanics innovation in agribusiness and address normal provokes and hindrances to execution. Furthermore, endeavors to advance limit building and innovation move in advanced mechanics schooling and preparing are fundamental for preparing ranchers and horticultural experts with the abilities and information they need to saddle the capability of advanced mechanics in agriculture.However, similarly, as with any troublesome innovation, the far-reaching reception of advanced mechanics in agribusiness additionally presents difficulties and potential dangers that should be tended to. Worries about information protection and security, licensed innovation freedoms, and administrative consistency should be painstakingly considered to guarantee that ranchers and partners are safeguarded and that advanced mechanics innovation is sent dependably and morally. Furthermore, endeavors to address the advanced separation and guarantee fair admittance to mechanical technology innovation for ranchers in emerging nations and underestimated networks are fundamental for advancing comprehensive and economical horticultural development.In the end, mechanical technology is changing agribusiness by offering ranchers new apparatuses and advancements to further develop efficiency, manageability, and flexibility in food creation. From accuracy cultivating and information-driven decision-production to inventive exploration and coordinated effort, the joining of mechanical technology into agribusiness is changing how we develop, reap, and oversee crops. As we keep on outfitting the force of mechanical technology in agribusiness, let us stay focused on advancing comprehensive and maintainable cultivating rehearses that benefit ranchers, buyers, and the climate the same.

Accuracy Cultivating and the Rural Transformation

Pushing the Modernized Difference in Agribusiness and Common Locales: The European Commission on Agribusiness highlighted the meaning of cutting-edge change in cultivation and nation areas. Current information and correspondence developments (ICTs) play a dire part in enabling farmers to work even more unequivocally, capably, and financially.

These advances also partner with creators and customers in new ways, offering more conspicuous choices and straightforwardness. Regardless, commonplace districts in Europe and Central Asia face moves in embracing new advances due to the fragile structure, moderateness, nonattendance of care, electronic capacities, and authoritative issues. To address this, the FAO Regional Office for Europe and Central Asia has shaped a broad nearby action expected to organize science, improvement, and mechanized approaches. Driving Factors for Rural Change: Natural change remembers changes in occupations, land use, and associations among metropolitan and commonplace districts. Key primary purposes incorporate Natural Factors: These affect family direct and flexible changes, propelling rural occupations and land use change starting around 1980. Work Land-Undertakings: The exchange of these viewpoints drives the improvement of metropolitan nation associations. Resource Flimsiness and Money-related Execution: Researchers have perceived a one-way causal association between resource unsteadiness and monetary execution. This highlights the meaning of supervising resources really for useful development Commonplace Metropolitan Change Challenges: Quick country metropolitan change processes impact matter streams, resource tasks, and natural framework working. Changes in people scattering along the country's metropolitan slant expect a basic part in framing these movements.

Chapter 9: Robotics in Disaster Response: Enhancing Safety and Rescue Operations

Despite catastrophic events, mishaps, and crises, mechanical technology has arisen as a basic device for upgrading security and proficiency in salvage and calamity reaction tasks. From search and salvage robots and automated flying vehicles (UAVs) to remotely worked vehicles (ROVs) and independent robots, advanced mechanics innovation is reforming how crisis responders survey harm, find survivors, and convey help in calamity-impacted regions. In this section, we will investigate the job of advanced mechanics in a fiasco reaction and its effect on improving security and salvage operations.At the front of mechanical technology in calamity reactions are search and salvage robots outfitted with sensors, cameras, and correspondence frameworks that empower them to explore dangerous conditions and find survivors caught in rubble, garbage, or fallen structures. These robots can get to bound spaces, shaky designs, and different regions that are blocked off or excessively risky for human heroes, giving ongoing situational mindfulness and working on the proficiency and adequacy of search and salvage operations.Furthermore, automated ethereal vehicles (UAVs) and drones are being utilized to overview debacle-impacted regions from a higher place, giving flying symbolism, 3D planning, and warm imaging information to assist crisis responders with surveying harm, distinguish perils, and focus on salvage endeavors. Drones outfitted with high-goal cameras and sensors can rapidly and proficiently overview huge areas of land, ocean, or metropolitan conditions, empowering responders to distinguish survivors, evaluate foundation harm, and plan departure courses in genuine time.Moreover, remotely worked vehicles (ROVs) and independent submerged vehicles (AUVs) are being sent in calamity reaction situations, for example, sea mishaps, oil slicks, and submerged search and salvage tasks. These submerged robots can explore submerged conditions, examine lowered designs, and gather information and tests from the seabed, giving important experiences into the degree of harm and ecological effect and illuminating decision-production by crisis responders and natural agencies.Additionally, advanced mechanics innovation is empowering the improvement of automated exoskeletons and wearable gadgets that upgrade the strength, perseverance, and portability of specialists on call in misfortune circumstances. These wearable advanced mechanics frameworks can help firemen, paramedics, and other crisis faculty in conveying weighty burdens, exploring unpleasant territory, and performing requesting assignments, lessening the gamble of injury and weariness and empowering responders to work all the more really in testing environments. Furthermore, mechanical technology is working with correspondence and coordination among crisis responders and organizations using automated ground vehicles (UGVs) and portable robots furnished with correspondence and systems administration capacities. These robots can act as portable correspondence centers, transferring messages, sending information, and planning reaction endeavors in regions with restricted or disturbed correspondence foundations. Furthermore, robots furnished with clinical supplies, water, and other fundamental assets can convey help to remote or difficult to arrive at areas, giving help to survivors and easing the weight on wrecked crisis services. However, as mechanical technology innovation keeps on propelling, it additionally brings up significant issues and difficulties connected with morals, security, and responsibility in misfortune reaction tasks. Worries about the moral utilization of advanced mechanics, including issues like information protection, reconnaissance, and the potential for unseen side-effects, should be painstakingly considered to guarantee that advanced mechanics innovation is sent capably and morally in catastrophe circumstances. Also, endeavors to lay out clear rules and conventions for the utilization of advanced mechanics in misfortune reaction, as well as preparing and limiting working for crisis responders, are fundamental for guaranteeing that advanced mechanics innovation is successfully coordinated into crisis the board frameworks and adds to positive results for survivors and networks impacted by disasters.In the end, advanced mechanics are altering debacle reactions by giving crisis responders new apparatuses and advances to improve security, proficiency, and viability in salvage tasks. From search and salvage robots and robots to submerged vehicles and wearable gadgets, advanced mechanics innovation is changing how we get ready for and answer calamities, saving lives and relieving the effect of crises on networks all over the planet. As we keep on outfitting the force of advanced mechanics in catastrophe reaction, let us stay focused on advancing moral and dependable utilization of innovation and guaranteeing that mechanical technology innovation helps all individuals, particularly those generally powerless against calamities and emergencies. Moreover, as advanced mechanics innovation keeps on developing, there is developing interest in investigating its likely applications in further developing fiasco readiness and flexibility in weak networks. Mechanical technology-empowered frameworks, for example, early advance notice frameworks, flood checking organizations, and avalanche location frameworks offer open doors for early recognition and reaction to normal perils, empowering networks to go to proactive lengths to lessen risk and moderate the effect of calamities. Moreover, advanced mechanics innovation can work with local area-based debacle readiness and reaction endeavors by enabling neighborhood occupants with the information and apparatuses they need to answer actually to crises and safeguard themselves and their communities. Furthermore, advanced mechanics is working with global coordinated effort and collaboration in catastrophe reaction through drives like the Worldwide Mechanical Technology Rivalry for Salvage Robots (RoboCup Salvage) and the DARPA Mechanical Technology Challenge. These rivalries unite groups of specialists, architects, and crisis responders from around the world to create and test mechanical frameworks for catastrophe reaction situations like quakes, fierce blazes, and atomic mishaps. By cultivating coordinated effort and information trade among partners, these contests speed up the turn of events and organization of advanced mechanics innovation in calamity reaction and add to further developed results for survivors and networks impacted by fiascos. Furthermore, advanced mechanics innovation is being coordinated into calamity reaction preparation and reproduction activities to improve the readiness and capacities of crisis responders. Computer-generated reality (VR) and expanded reality (AR) reenactments empower responders to rehearse and refine their abilities in reasonable fiasco situations, working on their capacity to successfully explore complex conditions, speak with colleagues, and settle on choices under tension. By giving vivid and intuitive preparation encounters, advanced mechanics empowered reproductions to assist crisis responders with building certainty and ability in misfortune reaction tasks, at last working on their status to answer true emergencies.Moreover, advanced mechanics innovation is empowering the improvement of independent and semi-independent frameworks for calamity reaction coordinated factors and production network the executives. Automated ground vehicles (UGVs) and aeronautical robots outfitted with freight conveyance frameworks can move fundamental supplies like food, water, clinical supplies, and safe house materials to calamity-impacted regions, even in remote or difficult-to-reach areas. These mechanical operations frameworks empower quick and productive conveyance of help to survivors and uprooted populaces, lessening reliance on conventional stock chains and working on the practicality and adequacy of calamity reaction efforts.However, likewise, with any troublesome innovation, the boundless reception of mechanical technology in a fiasco reaction additionally presents difficulties and potential dangers that should be tended to. Worries about interoperability, normalization, and similarity among various mechanical frameworks and stages should be addressed to guarantee consistent combination and coordination in multi-organization debacle reaction activities. Also, endeavors to address moral and lawful contemplations, like responsibility and responsibility for mechanical activities in calamity circumstances, are fundamental for advancing the capable and moral utilization of advanced mechanics innovation in crisis management.In the end, advanced mechanics are changing catastrophe reactions by furnishing crisis responders with new apparatuses and innovations to upgrade security, productivity, and adequacy in salvage tasks. From search and salvage robots and robots to strategies frameworks and preparing reenactments, advanced mechanics innovation is changing how we plan for and answer calamities, saving lives and alleviating the effect of crises on networks all over the planet. As we keep on bridling the force of advanced mechanics in a fiasco reaction, let us stay focused on advancing joint effort, development, and capable utilization of innovation to fabricate versatile and feasible networks that can endure and recuperate from catastrophes and crises.

Deploying Robots in Emergency Situations

Robots assume a critical part in crisis reaction situations, assisting specialists on call with exploring perilous conditions and relieving chances. Here are a few manners by which robots are conveyed in crisis circumstances: Search and Salvage Activities: Robots can explore through trash, unsteady designs, and other risky regions to look for survivors after catastrophes like seismic tremors or building breakdowns. They give basic aeronautical film and situational mindfulness, helping responders in rapidly evaluating what is going on and giving guidance Unsafe Material Taking care of: Robots can deal with hazardous substances, like poisonous synthetics or radioactive materials, diminishing the gamble to human responders. They can enter regions where it's risky for people, limiting the possibilities of injury or mischief. Remote Detecting and Information Assortment: Ethereal robots and ground robots gather information from catastrophe-stricken regions, assisting responders with pursuing informed choices. They catch pictures, recordings, and sensor information, giving significant bits of knowledge to crisis the executives. Correspondence and Coordination:

Robots can lay out correspondence networks in regions with disturbed frameworks. They transfer data between responders, further developing coordination during crises. Foundation Investigation and Harm Appraisal: Robots survey the state of structures, spans, and different designs after debacles. They distinguish primary harm, spills, or different dangers, permitting responders to focus on their endeavors. Operations and Backing: Robots help with coordinated factors, shipping supplies, clinical hardware, and different basics to impacted regions. They let loose human responders to zero in on basic undertakings while taking care of routine operations.

Chapter 10: The Morals of Advanced Mechanics: Tending to Moral and Social Ramifications

As mechanical technology and man-made brainpower (computer-based intelligence) innovations keep on progressing quickly, inquiries concerning their moral ramifications have become progressively unmistakable. From worries about work dislodging and algorithmic predisposition to issues of security, responsibility, and independence, the moral elements of advanced mechanics are perplexing and complex. In this section, we will investigate the moral difficulties and problems presented by mechanical technology and man-made intelligence, and examine systems for addressing them to advance dependable and moral turn of events and sending of these technologies.At the core of the moral discussion encompassing mechanical technology and computer-based intelligence is the subject of what these advancements will mean for human culture and individual prosperity.

As mechanization replaces human work in different enterprises, worries about work dislodging, financial disparity, and social disturbance have become more articulated. Additionally, the potential for simulated intelligence calculations to propagate or compound existing predispositions and separation, especially in regions, for example, recruiting, loaning, and law enforcement, brings up significant issues about reasonableness, equity, and value in the utilization of computer-based intelligence frameworks. Moreover, the rising joining of advanced mechanics and artificial intelligence into regular day-to-day existence raises worries about security, reconnaissance, and the disintegration of individual independence. As shrewd gadgets and independent frameworks gather and investigate immense measures of individual information, inquiries concerning assent, information proprietorship, and algorithmic straightforwardness become fundamental. Also, the utilization of computer-based intelligence-fueled observation frameworks openly spaces raises worries about common freedoms, common liberties, and the potential for misuse or abuse of these advancements by state-run administrations and other actors.Moreover, the sending of independent frameworks like self-driving vehicles, drones, and mechanical weapons brings up significant moral issues about responsibility, obligation, and the assignment of dynamic positions to machines. As independent frameworks settle on choices continuously without human intercession, inquiries concerning the moral organization, risk, and the portion of liability regarding the outcomes of their activities become progressively intricate. Also, the potential for independent frameworks to inflict damage or unseen side-effects, either through breakdowns, mistakes, or purposeful abuse, raises significant moral contemplations about hazard, security, and the moral plan and guideline of artificial intelligence and mechanical technology systems.However, as we wrestle with these moral difficulties, it is fundamental to perceive the likely advantages of advanced mechanics and man-made intelligence in tending to squeezing cultural difficulties and propelling human government assistance. From further developing medical care results and upgrading openness for individuals with handicaps to tending to environmental change and advancing maintainable turn of events, mechanical technology, and man-made intelligence offer open doors for advancement and advancement that can work on the personal satisfaction of individuals all over the planet.

Besides, endeavors to address the moral components of advanced mechanics and simulated intelligence require coordinated effort and commitment across numerous partners, including policymakers, analysts, industry pioneers, and common society associations. By cultivating discourse, straightforwardness, and responsibility in the turn of events and sending of mechanical technology and computer-based intelligence advancements, we can guarantee that these advances are lined up with human qualities and add to the benefit of all. Furthermore, endeavors to advance variety, incorporation, and value in the turn of events and utilization of mechanical technology and man-made intelligence are fundamental for tending to predisposition and separation and guaranteeing that these innovations benefit all individuals in society.In the end, the moral difficulties presented by advanced mechanics and man-made intelligence are mind-boggling and multi-layered, requiring cautious thought and smart consideration by all partners. From worries about work dislodging and algorithmic predisposition to inquiries of security, responsibility, and independence, the moral elements of advanced mechanics and man-made intelligence are fundamental to their turn of events and organization. By tending to these difficulties with uprightness, straightforwardness, and a guarantee to human qualities, we can guarantee that mechanical technology and simulated intelligence innovations add to an all the more, impartial, and maintainable future for all. Frameworks for guidelines and administration that guarantee the capable and moral turn of events, organization, and utilization of these advances. Administrative bodies and policymakers assume an essential part in laying out rules and norms for the moral plan and activity of mechanical technology and man-made intelligence frameworks, as well as observing consistency and implementing responsibility. Additionally, worldwide participation and cooperation are fundamental for orchestrating guidelines and standards across borders and advancing worldwide principles for the moral utilization of mechanical technology and AI. Additionally, endeavors to advance moral contemplations in mechanical technology and artificial intelligence should be coordinated into schooling and preparing programs for designers, engineers, and different experts engaged with the plan and execution of these advances. By integrating morals training into STEM educational plans and expert advancement programs, we can guarantee that people in the future of technologists are outfitted with the information and abilities they need to explore the moral intricacies of mechanical technology and man-made intelligence and settle on informed choices that focus on human government assistance and well-being.Moreover, cultivating public mindfulness and commitment to the moral ramifications of mechanical technology and man-made intelligence is fundamental for building trust and advancing capable stewardship of these innovations. Public exchange, resident interest, and partner commitment can assist with bringing issues to light about the expected dangers and advantages of mechanical technology and simulated intelligence, as well as enable people and networks to advocate for the moral and responsible utilization of these innovations. Furthermore, endeavors to advance straightforwardness and receptiveness in the turn of events and sending of mechanical technology and artificial intelligence can assist with building public trust and trust in these technologies.Furthermore, interdisciplinary exploration and joint effort are fundamental for propelling comprehension we might interpret the moral components of mechanical technology and man-made intelligence and create procedures for tending to moral difficulties and issues. By uniting specialists from different fields like way of thinking, morals, regulation, social science, and software engineering, we can encourage interdisciplinary discourse and coordinated effort that improves how we might interpret the moral ramifications of mechanical technology and man-made intelligence and illuminates morally independent direction and strategy development.Ultimately, tending to the moral ramifications of advanced mechanics and simulated intelligence requires a comprehensive and multi-layered approach that incorporates innovative advancement, administrative oversight, instruction and preparation, public commitment, and interdisciplinary exploration. By cooperating to address the moral difficulties and predicaments presented by mechanical technology and artificial intelligence, we can guarantee that these advances add to an all the more, fair, and reasonable future for all.In the end, the moral ramifications of advanced mechanics and artificial intelligence are significant and expansive, addressing basic inquiries regarding human qualities, freedoms, and obligations in an undeniably robotized and interconnected world. By tending to these difficulties with trustworthiness, straightforwardness, and a guarantee of human government assistance, we can tackle the extraordinary capability of mechanical technology and computer-based intelligence to make a morally solid future, socially, and earth economically for a long time into the future.

Balancing Innovation with Responsibility
To strike a balance between innovation and responsibility in robotics, ethical considerations must be taken into consideration at every stage of development and implementation. Ethical Design: When designing a robotic system, ethical considerations must be taken into account. It is necessary to employ developers who are moral and capable of incorporating responsibility into robotic technologies. Control versus freedom: As robots become more independent, it is essential to establish clear guidelines and control mechanisms to ensure ethical decision-making and prevent misuse. Data privacy and security Robots collect a lot of data, so privacy and security are important. This includes discussing the ethical implications of robotic systems handling data. Attribution of Responsibilities: The procedures for delegating responsibilities must be followed by all parties involved in the creation and operation of a robot. As a result, moral coherence and accountability are maintained. Ethical robotics promotes responsible behavior and emphasizes the well-being of workers. This includes taking into account the employment implications of people whose jobs involve interacting with robots. Transparency and the Elimination of Bias: To

ensure that robotics technologies are equitable and do not exacerbate the situation, it is necessary to Take Measures to Reduce Bias and Transparency in the Application of Artificial Intelligence to Robots. The ultimate goal is to guarantee that robotics technologies are developed and utilized in ways that enhance our lives, our safety, and society as a whole. You can read the articles on these topics for more in-depth insights.

Chapter 11: The effects of robots on employment on job and workforce dynamics

Discussions about the future of work and the potential impact on employment and workforce dynamics have arisen as a result of the incorporation of robotics and automation into various industries. The nature of jobs and the skills required for success in the workforce are being transformed by robotics technology in industries as diverse as manufacturing, logistics, healthcare, and service. One of the primary concerns surrounding the rise of robotics is the potential for job displacement and changes in the composition of the workforce. In this chapter, we will investigate the implications of robotics on employment, workforce dynamics, and strategies for navigating the changing landscape of work in the age of automation. Workers whose jobs are susceptible to automation run the risk of losing their jobs as routine and repetitive tasks are replaced by automation in the manufacturing and assembly industries. In addition, advancements in robotics technology, such as the creation of AI-powered systems and autonomous robots, may have an impact on white-collar professions like administrative work, data entry, and customer service in addition to traditional blue-collar jobs. On the other hand, although robotics technology may result in the loss of some jobs, it also opens up new opportunities for employment and economic expansion. New jobs in fields like software development, data analysis, system integration, robotics maintenance and repair, and automation can emerge as a result of their use. In addition, there is a growing demand for skilled workers who are capable of designing, operating, and managing robotic systems as well as interpreting data generated by these systems. In addition, the technology of robotics has the potential to boost productivity, efficiency, and competitiveness in industries that implement automation, which would result in an overall increase in employment and economic expansion. Robotics technology can free up human workers to concentrate on higher-value tasks that require creativity, critical thinking, and problem-solving abilities by automating routine and repetitive tasks. Additionally, robotics technology is driving the evolution of workforce dynamics and reshaping the skills required for success in the 21st-century labor market. Robots-enabled systems such as collaborative robots (cobots) can enhance human capabilities and enhance workplace safety by assisting workers with physically demanding tasks and reducing the risk of injuries and accidents. There is a growing demand for investments in education and training programs that equip individuals with the skills and competencies necessary to thrive in a technology-driven economy as the demand for workers with technical skills in robotics, programming, and data analysis grows. As automation alters the nature of work and how we collaborate and interact with machines and AI systems, soft skills like adaptability, communication, and teamwork become increasingly important. On the other hand, as we navigate the shifting landscape of work in the age of automation, it is essential to address concerns regarding equity, access, and inclusion in the workforce. To ensure that everyone has the chance to adapt to and thrive in the digital economy, efforts to promote lifelong learning and reskilling programs are crucial, particularly for workers who are at risk of losing their jobs due to automation. Diversity, equity, and inclusion in STEM education and workforce development are also essential for creating a workforce that reflects our society's diversity and uses robotics technology to its full potential for innovation and economic growth. In conclusion, the incorporation of robotics and automation into the workforce presents individuals, businesses, and society as a whole with both opportunities and challenges. While robotics technology has the potential to boost productivity, efficiency, and competitiveness, it also raises concerns about job displacement, skill gaps, and inequality in the workforce. We can ensure that robotics technology contributes to a future where work is meaningful, inclusive, and sustainable for all by proactively addressing these challenges through investments in education, training, and workforce development. Furthermore, efforts to mitigate the potential negative effects of robotics on employment require collaboration and coordination among stakeholders, including policymakers, businesses, educators, and labor organizations. Workforce training programs, apprenticeships, and job transition assistance are examples of policy interventions that can assist workers in acquiring the skills they need to succeed in a technology-driven economy and adapting to shifting job requirements. In addition, efforts to boost economic growth and job creation in industries that complement robotics and automation, such as digital services, renewable energy, and advanced manufacturing, can offset job losses in automation-affected industries. In addition, to take advantage of the business opportunities provided by robotics and automation, it is essential to cultivate a culture of innovation and entrepreneurship. Governments can boost innovation and open up new avenues for job creation and economic expansion by providing incentives for startups and small businesses, encouraging partnerships between academia and industry, and supporting research and development initiatives. Additionally, as robotics technology continues to advance, there is a growing need for ethical and responsible approaches to automation that place a priority on human well-being and social welfare. Additionally, efforts to promote the commercialization of robotics research and technology transfer can assist in translating scientific discoveries into practical applications that benefit society and contribute to economic prosperity. Ethical guidelines for robotics system design and implementation, transparency and accountability mechanisms for AI algorithms, and public participation in decision-making processes can all contribute to ensuring that robotics technology is developed and used by human values and for the benefit of the general public. In conclusion, the impact of robotics on employment and workforce dynamics is complex and multifaceted, with opportunities and challenges for individuals, businesses, and society as a whole. To build a future in which robotics technology benefits all members of society, efforts to address the social and economic implications of automation, such as income inequality, job polarization, and access to healthcare and social services, are essential. We can navigate the changing landscape of work in the age of automation and ensure that robotics technology contributes to a future where work is meaningful, inclusive, and sustainable for all by embracing innovation, investing in education and training, and fostering collaboration and dialogue among stakeholders.

Making Adjustments to the Changing Employment Landscape

Indeed, a crucial issue is adapting to the shifting employment landscape, particularly in light of the rise of robotics and automation. Consider these important points: Increased Automation: Contrary to popular belief, automation, and robotics are altering the nature of work rather than necessarily replacing workers. As businesses become more productive and competitive, increased automation may result in an overall increase in hiring. Management changes: The introduction of robots may reduce the need for managers, particularly those in charge of high-skilled employees. This is because robots can reduce human error and increase efficiency. Upskilling and Reskilling: Workers need to be upskilled and reskilled to be able to adapt to new technologies. Repetitive or straightforward problem-solving tasks are the most susceptible to automation. Collaboration between humans and AI: The key is to foster a culture of continuous learning and acknowledge the significance of human abilities. It will be essential to adjust to a hybrid workforce in which AI and humans collaborate. New jobs are being created, even though automation may eliminate some jobs. However, new roles are being created that require different skill sets. In conclusion, the focus ought to be on leveraging technology to become more productive and competitive while also ensuring that workers are prepared for the changes brought about by robotics and automation. It is important to ensure that workers are equipped with the skills necessary to fill these new roles. It's about striking a balance between human labor and technological advancements.

Chapter 12: Accessibility and robotics: Giving people with disabilities more power

The way people with disabilities interact with their environment has been transformed by the incorporation of robotics technology into assistive devices and accessibility solutions, enhancing their independence, mobility, and quality of life. Assistive robots, smart home systems, robotic prosthetics, and exoskeletons are just a few examples of how robotics technology is enabling people with disabilities to overcome physical barriers and fully participate in society. Robotic prosthetics and exoskeletons are transforming the lives of people with limb loss or mobility impairments by restoring mobility, dexterity, and functionality. In this chapter, we will examine the role of robotics in accessibility and its effect on empowering people with disabilities. Prosthetic limbs with AI algorithms, sensors, and actuators can mimic the natural movements of human limbs, making it easier and more precise for users to carry out a wide range of everyday tasks. Additionally, robotics technology is facilitating the development of assistive robots and robotic companions that support and assist people with disabilities in a variety of aspects of daily life. In a similar vein, exoskeletons and powered orthotic devices can assist people with mobility impairments by providing support and assistance with walking, standing, and climbing stairs. This enables the individuals to navigate their environment with greater independence and confidence. Social robots with AI and natural language processing capabilities can help people with disabilities feel less lonely and isolated by helping with things like communication, social interaction, and emotional support. Further, robotics technology is revolutionizing accessibility in the built environment by enabling the development of smart home systems and environmental control devices that are tailored to the needs of individuals with disabilities.

Additionally, service robots with manipulators and sensors can assist with tasks such as personal care, meal preparation, and household chores, allowing individuals with disabilities to live more independently and autonomously. People with disabilities can live more comfortably and safely in their own homes thanks to smart home systems that are equipped with sensors, actuators, and voice recognition technology. These systems can automate and control various aspects of the home environment, such as lighting, temperature, and security. Additionally, the development of accessible transportation systems, communication devices, and assistive technologies is facilitating access for people with disabilities to education, employment, and social participation. Additionally, environmental control devices such as adaptive switches, voice-activated assistants, and gesture recognition systems enable individuals with disabilities to control electronic devices and appliances with greater ease and independence. People with mobility impairments can travel safely and independently thanks to autonomous vehicles that are equipped with wheelchair-accessible features and assistive technologies. As a result, barriers to employment, education, and community participation are reduced. Similarly, speech-generating devices, braille displays, and alternative input devices enable people with communication disabilities to express themselves and interact with others more effectively, fostering inclusion and participation in society. On the other hand, although robotics technology has the potential to transform the lives of people with disabilities, it also raises significant concerns regarding accessibility, affordability, and usability. To ensure that all people with disabilities have equal access to robotics-enabled assistive technologies, it is necessary to address concerns regarding the cost and availability of these devices, as well as the requirement for training and support for users and caregivers.

In conclusion, robotics technology is revolutionizing accessibility by providing innovative solutions that empower people with disabilities to overcome physical barriers and participate more fully in society. Additionally, efforts to address ethical and social considerations, such as privacy, autonomy, and the potential for dependence on technology, are essential for promoting responsible and ethical use of robotics technology in accessibility solutions. Robotics technology is improving the independence, mobility, and quality of life of people with disabilities through assistive robots, exoskeletons, robotic prosthetics, and smart home systems. To ensure that robotics technology benefits all members of society, regardless of ability or disability, let us keep our commitment to promoting inclusive design, equitable access, and ethical technology use as we continue to harness robotics' accessibility potential. Additionally, to advance accessibility in robotics, efforts to foster collaboration and partnership among stakeholders, such as researchers, engineers, healthcare professionals, policymakers, and advocacy organizations, are essential. We can ensure that assistive technology and accessibility solutions meet the diverse needs and preferences of people with disabilities by accelerating innovation and development by fostering interdisciplinary collaboration and knowledge exchange. The public's understanding and support for robotics technologies are also dependent on efforts to raise education and awareness about robotics technology and accessibility. We can encourage acceptance and adoption of assistive technologies among people with disabilities, caregivers, and the general public by promoting awareness of the potential accessibility benefits of robotics and dispelling misconceptions. In addition, it is essential to address regulatory and policy barriers to the development and deployment of robotics technology in accessibility to ensure equitable access and adoption of these technologies. Additionally, it is essential to empower people with disabilities to use robotics-enabled assistive devices effectively and independently. Policy interventions like funding incentives, procurement policies, and accessibility standards can encourage investment in robotics-enabled assistive device research and development and guarantee that these technologies satisfy the requirements of people with disabilities. In conclusion, robotics technology has the potential to transform the lives of people with disabilities by providing innovative solutions that enhance independence, mobility, and quality of life. In addition, efforts to promote universal design principles and accessibility standards in the development of robotics technology are essential for ensuring that these technologies are usable and accessible to people with diverse abilities and disabilities. Robotics-enabled assistive devices, such as assistive robots, smart home systems, and robotic prosthetics and exoskeletons, are enabling people with disabilities to overcome physical barriers and participate more fully in society. To ensure that robotics technology benefits all members of society, regardless of ability or disability, let us keep our commitment to promoting inclusive design, equitable access, and ethical technology use as we continue to advance the accessibility of robotics.

Enhancing Accessibility through Assistive Robotics

Indeed, a crucial issue is adapting to the shifting employment landscape, particularly in light of the rise of robotics and automation.

Consider these important points: More automation: Contrary to popular belief, automation, and robotics are changing the nature of work rather than replacing workers. As businesses become more productive and competitive, increased automation may result in an overall increase in hiring. Management changes: The introduction of robots may reduce the need for managers, particularly those in charge of high-skilled employees. This is because robots can reduce human error and increase efficiency. Upskilling and Reskilling: Workers need to be upskilled and reskilled to be able to adapt to new technologies. Repetitive or straightforward problem-solving tasks are the most susceptible to automation.

> ➢ Human-AI Collaboration: It is essential to foster a culture of continuous learning and acknowledge the significance of human abilities. It will be essential to adjust to a hybrid workforce in which AI and humans collaborate. The creation of new jobs: Although automation may

eliminate some jobs, new roles are being created that call for different skill sets. It's critical to make sure workers have the skills they need to fill these new positions. In a nutshell, improving productivity and competitiveness through the use of technology should be the primary focus, as should preparing employees for the changes brought about by automation and robotics. It's about striking a balance between human labor and technological advancements.

Chapter 13: Exploring the Limits of Creativity Through the Use of Robots in Entertainment

The way we experience and interact with entertainment media has been transformed by the introduction of robotics technology into the entertainment industry, heralding a new era of creativity and innovation. Robotics-enabled attractions and experiences are captivated by audiences and pushing the boundaries of storytelling and immersive entertainment in everything from theme parks to live performances to film, television, and gaming. One of the most obvious manifestations of robotics in entertainment is found in theme parks and attractions, where animatronics and robotic characters bring fantasy worlds to life and create immersive experiences for visitors. In this chapter, we will investigate the role of robotics in entertainment and its impact on shaping the future of the entertainment industry. The creation of dynamic and engaging environments that transport guests to fantastical worlds and pique their imaginations is made possible by robotics technology, which enables theme park designers and Imagineers to create lifelike dinosaurs, creatures, interactive robots, and animatronic figures. Additionally, advancements in robotics technology, such as the use of sensors, actuators, and artificial intelligence (AI) algorithms, are making it possible for attractions at theme parks to become more interactive and responsive to the input of guests, which is improving the overall entertainment experience. Additionally, robotics technology is revolutionizing theatrical productions and live performances by enabling the development of dynamic and expressive robotic characters and performers. With mesmerizing displays of movement, expression, and emotion, robotics-enabled performances push the boundaries of what is possible in live entertainment with puppetry, kinetic sculptures, robotic actors, and dancers. They also blur the line between humans and machines. In addition, robotics technology is transforming the film and television industry by enabling filmmakers and content creators to bring imaginary worlds and characters to life with unprecedented realism and detail. By leveraging the capabilities of robots, robotics technology enables performers and artists to explore new forms of expression and storytelling. Robotics technology enables filmmakers to create immersive and believable worlds that captivate audiences and elicit powerful emotional responses, from animatronic creatures and robotic props to characters and visual effects enhanced with computer-generated imagery (CGI). In addition, robotics technology is reshaping the gaming landscape by enabling the creation of immersive and interactive experiences that blur the boundaries between the virtual and physical worlds. In addition, robotics technology is reshaping the landscape of gaming by enabling the development of immersive and interactive experiences that blur the boundaries between the virtual and physical worlds. By providing tactile feedback, haptic sensations, and physical interaction with virtual environments, robotics technology enhances gameplay and immersion, from robotic gaming peripherals and accessories to augmented reality (AR) and virtual reality (VR) experiences. Additionally, robotics-enabled gaming experiences provide players with opportunities to engage with games in novel and exciting ways, such as through motion-controlled interfaces, gesture recognition, or voice commands. On the other hand, as robotics technology progresses and becomes increasingly ingrained in entertainment media, it also raises significant concerns regarding ethics, safety, and the future of employment in the entertainment industry. To ensure that robotics-enabled experiences are inclusive, respectful, and culturally sensitive, it is necessary to carefully consider ethical concerns regarding the use of robotics in entertainment, such as consent, privacy, and representation. In conclusion, robotics technology is revolutionizing the entertainment industry by pushing the boundaries of creativity and imagination and creating new opportunities for immersive and interactive experiences. To guarantee the safe operation of robotics-enabled attractions and experiences in entertainment venues, efforts to address safety considerations such as risk assessment, emergency protocols, and user training are essential. Robotics-enabled attractions and experiences are captivating audiences and transforming the way we experience and engage with entertainment media, ranging from theme parks to live performances, films, television, and video games. It is essential to foster collaboration and innovation between robotics engineers, professionals in the entertainment industry, and creative artists to drive the development of cutting-edge robotics-enabled entertainment experiences as we continue to investigate the intersection of technology and imagination in entertainment. Let us remain committed to promoting the ethical and responsible use of robotics technology and ensuring that robotics-enabled experiences enrich and inspire audiences worldwide. In addition, robotics technology is democratizing access to entertainment creation and consumption by enabling individuals and communities to participate in the production and distribution of content. We can push the boundaries of what is possible in entertainment by bringing together expertise from diverse fields like robotics, engineering, animation, storytelling, and design. Technology in robotics gives enthusiasts and creators the ability to experiment with robotics and create their own interactive experiences and content through online platforms, social media, maker communities, and DIY robotics kits. Further, robotics technology is driving innovation in entertainment marketing and promotion by enabling the development of interactive and engaging experiences that capture audience attention and drive brand engagement. In addition, robotics-enabled tools and platforms for content creation and distribution enable creators to reach global audiences and share their creations with the world, democratizing access to entertainment and fostering creativity and innovation in the digital age. Brands and advertisers can use robotics technology to create memorable and shareable experiences that resonate with consumers and cultivate brand loyalty. These experiences can include immersive installations, experiential marketing campaigns, robotic mascots, and characters. In addition, robotics-enabled retail experiences like interactive displays and robotic product demonstrations improve the shopping experience and increase customer engagement and sales. On the other hand, as robotics technology continues to revolutionize the entertainment industry, it also raises significant concerns regarding privacy, security, and the ethical use of technology. To ensure that audience rights and interests are protected, robotics-enabled entertainment experiences must address concerns regarding data privacy, surveillance, and the collection and use of personal information. In conclusion, robotics technology is transforming the entertainment industry by pushing the boundaries of creativity and imagination and creating new opportunities for immersive and interactive experiences. Additionally, efforts to address safety considerations, such as risk assessment, regulatory compliance, and user education, are essential for ensuring the safe operation of robotics-enabled attractions and experiences and minimizing the risk of accidents or injuries. Robotics-enabled attractions and experiences are captivating audiences and reshaping the way we experience and engage with entertainment media. They can be found in everything from theme parks to live performances to film, television, gaming, and marketing. Let us keep our commitment to promoting the ethical and responsible use of technology and ensuring that robotics-enabled experiences enrich and inspire audiences worldwide as we continue to harness the power of robotics in entertainment.

From Animatronics to Interactive Performers

A significant development in the entertainment and robotics industries can be seen in the shift from animatronics to interactive performer robotics. A summary of this transformation is as follows: The traditional meaning of "animatronics" is "the use of mechanical devices to animate robotic figures," which are frequently found in motion pictures, amusement parks, and other entertainment venues.

These figures can copy similar developments however are normally restricted to pre-customized activities. In contrast, Interactive Performer Robotics creates robots that can interact with humans and their surroundings in real-time by incorporating cutting-edge technologies like sensors, cameras, and artificial intelligence. Because of this, the performance can be more dynamic and adaptable, with the robot being able to respond to the audience or changes in the environment123. For instance, animatronic figures in theme parks provide lifelike movements; however, the incorporation of robotics has made these attractions much more adaptable, enabling content to be reprogrammed and updated on the fly. Robots are currently being developed for use in social applications, such as education, entertainment, or assisted living, outside of the realm of entertainment. The new method of character animation known as animation in robotics extends the traditional method by enabling animated motion to become more interactive and adaptable during user interaction in real-world settings. Artists and developers of robots work together to develop expressive, emotional, and design characteristics for robots that can meaningfully interact with people. Overall, the move toward interactive performer robotics, in which robots are both performers and participants in the interaction, indicates a move toward creating entertainment experiences that are more immersive and engaging.

Chapter 14 Understanding the complexities of military applications through robotics and warfare

The landscape of contemporary warfare and security has been transformed by the incorporation of robotics technology into military applications. As a result, new capabilities and difficulties have emerged for both military forces and policymakers. Robotics technology is changing the way military operations are done and bringing up important ethical, legal, and strategic issues.

These include autonomous weapons systems, ground robots, and surveillance drones. Unmanned aerial vehicles (UAVs), more commonly referred to as drones, have become increasingly prevalent in military operations for reconnaissance, surveillance, and targeted strikes. In this chapter, we will examine the complexities and implications of its military applications, as well as the role that robotics plays in security and warfare. While surveillance drones provide commanders on the ground with real-time intelligence and situational awareness, armed drones with precision-guided munitions enable military forces to carry out surgical strikes against enemy targets with the least amount of risk to personnel and collateral damage. In addition, robotics technology is revolutionizing ground warfare through the development of unmanned ground vehicles (UGVs) and robotic systems for reconnaissance, surveillance, and combat support. In addition, advancements in autonomy and AI algorithms are enabling drones to operate autonomously and collaboratively in swarms, enhancing their effectiveness and versatility in a wide range of military missions. UGVs with sensors, cameras, and manipulators can traverse obstacles, navigate rough terrain, and perform a variety of tasks like mine clearance, route clearance, and explosive ordnance disposal (EOD). This makes military operations safer and more efficient. Additionally, robotics technology is driving innovation in naval warfare through the development of unmanned surface vessels (USVs) and underwater drones for maritime surveillance, mine countermeasures, and anti-submarine warfare. Additionally, robotic systems like robotic exoskeletons and unmanned combat vehicles (UCVs) enable soldiers to enhance their capabilities and overcome physical limitations on the battlefield, enhancing their mobility, endurance, and lethality in combat. Enhancing maritime security and defense capabilities, USVs equipped with sensors, sonar, and communication systems can autonomously patrol maritime borders, monitor shipping lanes, and identify and neutralize underwater threats. As robotics technology continues to advance and become more integrated into military operations, it also raises important ethical, legal, and strategic considerations that must be carefully addressed. Additionally, underwater drones equipped with cameras and sensors enable naval forces to conduct underwater reconnaissance, search and rescue operations, and environmental monitoring in underwater environments that are hazardous or inaccessible to manned vehicles. To ensure that robotics-enabled warfare is carried out in a manner that is by human rights and ethical principles, it is necessary to carefully consider ethical concerns regarding the use of autonomous weapons systems. These concerns include issues such as accountability, transparency, and compliance with international humanitarian law (IHL). In conclusion, robotics technology is reshaping the landscape of modern warfare and security, introducing new capabilities and challenges for military forces and policymakers alike. To promote stability and security in an increasingly complex and contested security environment, efforts to address the strategic implications of robotics technology, such as arms races, proliferation, and escalation dynamics, are essential. Robotics technology is changing the way military operations are done and bringing up important ethical, legal, and strategic issues in everything from unmanned aerial vehicles and ground robots to autonomous weapons systems and underwater drones. The efforts to address the ethical, legal, and strategic implications of robotics in warfare necessitate collaboration and coordination among military leaders, policymakers, ethicists, legal experts, and civil society organizations. Let us keep our commitment to promoting the responsible and ethical use of technology and ensuring that robotics-enabled military applications contribute to peace, security, and stability in the international system as we continue to navigate the complexities of robotics in warfare. The development of norms, guidelines, and regulations that govern the development, deployment, and use of robotics-enabled military technologies as well as the observance of international law and human rights standards require international dialogue and cooperation. In addition, efforts to promote responsible innovation and risk management in the development and deployment of robotics-enabled military technologies are essential for ensuring the safety, dependability, and effectiveness of these systems. Furthermore, efforts to promote transparency, accountability, and oversight mechanisms for robotics-enabled military operations are essential for building trust and confidence among stakeholders and minimizing the risk of unintended consequences or misuse of these technologies. To evaluate the performance and dependability of robotics-enabled military technologies under a variety of operational conditions and to identify and mitigate potential risks and vulnerabilities, robust testing, evaluation, and validation procedures are required. Additionally, efforts to promote human-machine collaboration and decision-making in warfare are essential for leveraging the strengths of both humans and machines while mitigating the limitations and risks of autonomous systems. Additionally, efforts to address cybersecurity threats and vulnerabilities in robotics-enabled military systems are essential for protecting against unauthorized access, tampering, or exploitation of these technologies by adversaries. For autonomous weapons systems to operate by human values and ethical principles and to prevent unintended harm or misuse, human oversight, and control mechanisms are necessary. The integration of robotics technology into military applications is reshaping the landscape of modern warfare and security, introducing new capabilities and challenges for military forces and policymakers alike. Additionally, efforts to promote human-machine teaming and collaboration, such as training and education programs for military personnel, are essential for enhancing the effectiveness and resilience of military forces in an operational environment that is becoming increasingly complex and dynamic. Robotics technology is changing the way military operations are done and bringing up important ethical, legal, and strategic issues in everything from unmanned aerial vehicles and ground robots to autonomous weapons systems and underwater drones. Let us keep our commitment to promoting the responsible and ethical use of technology and ensuring that robotics-enabled military applications contribute to peace, security, and stability in the international system as we continue to navigate the complexities of robotics in warfare.

Analyzing Robotics' Contribution to Defense Strategies

Because it provides a variety of capabilities that enhance military operations, robotics has become an essential component of contemporary defense strategies. The following are some significant contributions of robotics to defense: Enhanced Surveillance and Reconnaissance: The technology behind robotics has made it much easier to carry out surveillance and reconnaissance. These missions now make use of real-time data and intelligence gathered from far-off or risky locations. Combat and precision strikes are made possible by unmanned systems like drones, which reduce the risk to military personnel. While minimizing collateral damage, they can engage targets with high accuracy. Management of the Logistics and Supply Chain Using robots, logistics, and supply chain operations can be streamlined to ensure that troops in the field effectively receive supplies and equipment. Explosive Ordnance Disposal (EOD): Robots are frequently used for EOD tasks because they make it possible to safely identify and eliminate explosive threats without putting lives at risk. Disaster Relief and Humanitarian Assistance: Robots can provide aid and support in disaster-stricken areas where humans may be too risky to operate. This can be an important part of humanitarian missions. Autonomous Vehicles and Unmanned Tanks: The development of autonomous vehicles and unmanned tanks is reshaping the battlefield, providing new tactical options, and decreasing the requirement for human soldiers in direct combat. Ethical and Legal Issues: The rise of military robotics raises several ethical and legal issues as well. These issues include the necessity of clear rules of engagement and the use of lethal autonomous weapons systems. As nations navigate the complexities of this rapidly advancing technology, the proliferation of robotics in the military has implications for international relations and arms control. The three elements of goals means, and threats are taken into account in the military strategic view of robotics. It emphasizes the significance of incorporating robotics into military education and training2 and the necessity of political, strategic, operational, and tactical levels of planning. You can refer to scholarly articles and reports that discuss the strategic implications of robotics in military contexts for a more in-depth analysis. Robotics and autonomous systems (RAS) will be crucial for the development of future military capabilities as they continue to evolve.

Chapter 15: From Companionship to Coexistence: The Direction of human-robot Interaction in the Future

The future of human-robot interaction holds enormous promise for transforming the way we live, work, and interact with technology, as robotics technology continues to advance. Robots have the potential to play increasingly important roles in our day-to-day lives, from being companions and caregivers to collaborating with humans in a variety of fields.

One of the most intriguing aspects of the future of human-robot interaction is the potential for robots to serve as companions and caregivers for humans, particularly in contexts such as healthcare, eldercare, and mental health support. In this chapter, we will examine the evolving landscape of human-robot interaction and the potential for humans and robots to coexist harmoniously in society. Social robots equipped with natural language processing, emotional recognition, and empathy algorithms have made it possible for robots to interact with humans in more natural and intuitive ways, enabling them to provide companionship, assistance, and emotional support to those in need. Additionally, robots are increasingly being incorporated into a variety of aspects of daily life, from personal assistance and entertainment to household chores and errands, which can assist in addressing social isolation and loneliness among vulnerable populations like the elderly and people with disabilities. Smart devices and robotic assistants with AI and automation features can streamline daily routines, manage tasks and schedules, and increase productivity and efficiency at home and work. In addition, the future of human-robot interaction holds promise for collaboration and coexistence between humans and robots in various domains, including industry, education, and research. Furthermore, we can anticipate a proliferation of robotics-enabled services and applications in areas such as retail, hospitality, transportation, and customer service, transforming the way we interact with technology and access goods and services. With sensors and artificial intelligence (AI) algorithms, collaborative robots (cobots) can collaborate with humans in manufacturing, logistics, and other industrial settings to boost productivity and safety at work. As humans and robots increasingly interact and coexist in society, it is essential to address important considerations related to ethics, privacy, and societal impact. Additionally, robots are increasingly being used in educational settings to support learning and skill development, providing students in STEM subjects and other disciplines with interactive and hands-on experiences. To ensure that robotics technology is developed and utilized in a manner that is consistent with human values and ethical principles, it is necessary to carefully consider concerns regarding the ethical use of robots in various contexts, such as autonomy, accountability, and transparency. In conclusion, the future of human-robot interaction holds enormous potential for transforming the way we live, work, and interact with technology. Additionally, efforts to address privacy concerns such as data security, surveillance, and consent are essential for protecting individual rights. Robots have the potential to play increasingly important roles in our day-to-day lives, from being companions and caregivers to collaborating with humans in a variety of fields. In addition, efforts to promote inclusivity and accessibility in human-robot interaction are essential for ensuring that robotics technology benefits all members of society, regardless of age, ability, or background. Let us remain committed to promoting responsible and ethical use of technology and ensuring that humans and robots can coexist harmoniously in society as we continue to explore the possibilities of human-robot interaction. Promoting equitable access to and participation in human-robot interactions necessitates the creation of robots and interfaces that are comprehensible, user-friendly, and accessible to people with a variety of requirements and preferences. In addition, fostering a culture of responsible innovation and stewardship in the development and deployment of robotics technology is essential for addressing societal concerns and ensuring that the benefits of human-robot interaction outweigh the risks and challenges. Further, efforts to address disparities in access to robotics technology, such as affordability, availability, and digital literacy, are essential for ensuring that all individuals have the opportunity to benefit from the potential of robotics technology to enhance their lives and well-being. To identify and address ethical, legal, and social considerations associated with human-robot interaction, stakeholders—such as researchers, engineers, policymakers, ethicists, and civil society organizations—must collaborate and communicate with one another. In addition, it is essential to establish frameworks for governance and regulation that guarantee the responsible and ethical use of robotics technology as humans and robots increasingly interact and collaborate in various domains. This is because efforts to engage the public in discussions about the implications of robotics technology and empower individuals to participate in decision-making processes are essential for promoting transparency, accountability, and trust in the development and use of robotics technology. The development, deployment, and use of robotics technology are governed by guidelines, standards, and policies that address important considerations like safety, privacy, and liability. Regulatory bodies and policymakers play a crucial role in this process. In conclusion, the future of human-robot interaction holds enormous promise for transforming the way we live, work, and interact with technology. Harmonizing regulations and norms across borders and promoting global standards for the ethical use of robotics technology require international cooperation and collaboration. Robots have the potential to play increasingly important roles in our day-to-day lives, from being companions and caregivers to collaborating with humans in a variety of fields. Let us remain committed to promoting the responsible and ethical use of technology and ensuring that humans and robots can coexist harmoniously in society, enriching our lives and advancing our shared goals of progress and well-being as we continue to investigate the possibilities of human-robot interaction.

Analyzing Relationship Dynamics Between People and Robots

Human-robot interaction (HRI) encompasses several fascinating and intricate aspects of the dynamics of human-robot relationships. Understanding how humans perceive, interact with, and relate to robots in various contexts is at the heart of this interdisciplinary field. When analyzing these dynamics, the following are some important considerations that researchers make: Anthropomorphism is the idea that robots have human characteristics.

A robot's level of anthropomorphism can have a significant impact on how people interact with it. The term "assistive robotics" (AR) refers to robots that are made to help humans in a variety of ways, such as with their physical, social, mental, and emotional well-being. The dynamics of the relationship may be affected by these robots' performance in their roles. Autonomy: People's level of trust and reliance in robotic systems can be affected by a robot's level of autonomy, or its capacity to operate independently. Benchmarks: For the development of HRI, it is essential to establish standards for robot performance, safety, and ethical considerations. Embodiment: Since robots are real-world objects, their design and shape can influence how people interact with them. The physiological measure known as the galvanic skin response (GSR) can be utilized to evaluate the emotional state of a person interacting with a robot and provide insights into the dynamics of the relationship. Human-Computer Interaction (HCI): While HRI specifically examines the dynamics between humans and physically embodied robots1, HCI focuses on the interaction between humans and computers. Socially Assistive Robotics (SAR): This field studies robots that help people by interacting with them socially rather than physically. This can be important for elderly care and education. The term "socially interactive robots" (SIR) refers to robots that interact with humans through social interactions, such as communicating with them, expressing their feelings, and learning their social cues. The goal of the research is to develop human behavior models that can anticipate and enhance interactions with robots. For HRI to be successful, these models must be accurate and complete to guarantee safety, performance, and employee satisfaction. Studies also show that people develop stronger bonds with the robots they control, which may have an impact on how semi-autonomous robots are made and how well they work. In conclusion, to improve the design and interaction of robotic systems with humans, a multidisciplinary approach that takes into account psychological, sociological, and technological factors is required for analyzing the dynamics of the relationships between humans and robots.

Chapter 16: Mechanical Technology and Ecological Preservation: Safeguarding Nature with Innovative Arrangements

For addressing urgent ecological issues and protecting the natural world for future generations, incorporating robotics technology into environmental conservation efforts is a promising option. Innovative solutions for sustainable environmental management are provided by robotics technology, which includes reducing pollution and preventing habitat destruction as well as monitoring ecosystems and wildlife.

One of the primary applications of robotics technology in environmental conservation is in the monitoring and management of ecosystems and wildlife habitats. In this chapter, we will examine the role of robotics in environmental conservation and the potential for technological solutions to contribute to the preservation of nature. Natural landscapes can be surveyed and mapped with the help of unmanned aerial vehicles (UAVs) that are outfitted with cameras, sensors, and remote sensing technologies. They can also be used to monitor changes in vegetation and wildlife populations. In addition, underwater drones and autonomous underwater vehicles (AUVs) enable researchers to explore and monitor marine ecosystems, evaluate coral reefs, and study underwater biodiversity in inaccessible locations. Additionally, robotics technology is reshaping the process by which environmental data is gathered and analyzed, making it possible for researchers to collect large quantities of high-quality data in a manner that is both more effective and precise than ever before. Real-time data on ecosystem health and environmental conditions can be provided by autonomous environmental monitoring stations outfitted with sensors for measuring air and water quality, temperature, humidity, and other environmental parameters. This makes it possible to detect pollution, habitat degradation, and other threats to biodiversity earlier. Additionally, robotics technology is being utilized in the fight against environmental pollution and habitat destruction, providing innovative solutions for cleaning up contaminated sites, mitigating the effects of oil spills, and restoring degraded ecosystems. Furthermore, AI-powered data analysis algorithms are capable of processing and analyzing vast amounts of environmental data, identifying patterns, trends, and anomalies that can inform conservation strategies and decision-making. It is possible to use robotic systems, such as drones and unmanned ground vehicles (UGVs) with sensors and sampling tools, to find and monitor sources of pollution, evaluate environmental damage, and collect samples for analysis and remediation. In addition, reforestation and revegetation efforts in areas affected by deforestation, wildfires, and land degradation are made possible by robotic platforms for habitat restoration, such as autonomous seed dispersal systems and planting drones. Even though robotics technology holds great promise for environmental conservation, it also raises important questions and challenges regarding ethics, governance, and the unintended consequences of technological interventions. To ensure that technological solutions respect human rights and cultural values and contribute to outcomes that are equitable and sustainable, it is necessary to carefully consider concerns about the ethical use of robotics in environmental conservation, including concerns about privacy, autonomy, and the rights of indigenous communities. In conclusion, robotics technology has the potential to revolutionize environmental conservation efforts by providing innovative solutions for monitoring, managing, and restoring ecosystems and wildlife habitats. To promote the responsible and ethical use of robotics technology in environmental conservation, efforts to address regulatory and policy challenges such as liability, accountability, and intellectual property rights are essential. New opportunities for sustainable environmental management are provided by robotics-enabled environmental conservation technologies, which range from surveying landscapes and monitoring biodiversity to cleaning up pollution and restoring degraded ecosystems. In addition, efforts to promote collaboration and partnership between stakeholders, including researchers, conservationists, policymakers, local communities, and technology developers, are essential for maximizing the impact of robotics technology in environmental conservation. Let us keep our commitment to promoting the responsible and ethical use of technology and ensuring that technological solutions contribute to the preservation of nature and the well-being of current and future generations. We can develop and implement robotics-enabled conservation strategies that are contextually relevant, culturally sensitive, and socially inclusive by fostering interdisciplinary collaboration and knowledge exchange. In addition, efforts to promote innovation and entrepreneurship in the development and deployment of robotics technology for environmental conservation are essential for unlocking new opportunities and scaling up successful initiatives. Furthermore, efforts to engage and empower local communities in conservation efforts, such as citizen science initiatives and participatory monitoring, are essential for building community ownership and support for conservation goals and ensuring the long-term sustainability of conservation interventions. Incentives, grants, and prizes for robotics research and environmental conservation innovation can encourage investment in promising technologies and solutions and stimulate creativity. In addition, efforts to address capacity-building and technology transfer challenges in the adoption and deployment of robotics technology for environmental conservation are essential for ensuring that technological solutions reach those who need them most. Additionally, initiatives to promote the commercialization of robotics research and technology transfer can facilitate the translation of scientific discoveries into practical applications that benefit society and contribute to environmental sustainability. The capacity to use robotics technology effectively in conservation activities can be built through training and education programs for conservation practitioners, technicians, and local communities. The adoption and adaptation of robotics technology in various environmental contexts and regions can also be facilitated by technology transfer initiatives like partnerships between research institutions, technology developers, and conservation organizations. Additionally, public awareness and engagement in robotics-enabled environmental conservation are essential for gaining support and momentum for conservation goals and initiatives. Campaigns of outreach and communication that highlight the role of robotics technology in conservation success stories, highlight innovative solutions and best practices and involve the general public in activities related to citizen science and conservation can raise awareness of environmental issues and motivate action and participation. Additionally, robotics technology has the potential to revolutionize environmental conservation efforts by providing novel solutions for monitoring, managing, and restoring ecosystems and wildlife habitats. In conclusion, robotics technology has the potential to revolutionize environmental conservation efforts by providing innovative solutions for monitoring, managing, and restoring ecosystems and wildlife habitats. New opportunities for sustainable environmental management are provided by robotics-enabled environmental conservation technologies, which range from surveying landscapes and monitoring biodiversity to cleaning up pollution and restoring degraded ecosystems. Let us maintain our commitment to promoting the responsible and ethical use of technology and ensuring that technological solutions contribute to the preservation of nature and the well-being of current and future generations as we continue to use robotics to conserve the environment.

Utilizing Robots for Conservation Activities

Conservation efforts are increasingly incorporating the use of robots to address a variety of environmental issues. An overview of how robots are assisting conservation efforts can be found here: Monitoring of species and Data Collection Data collection on species and habitats is being transformed by robots, particularly drones and autonomous underwater vehicles (AUVs). They are capable of navigating difficult and remote terrains and collecting data on species populations, health, and behavior without human intervention, which is essential for delicate ecosystems.

Contribution to Pollination Robotic pollinators has been developed in response to the decline of natural pollinators like bees. To maintain plant populations and genetic diversity within ecosystems, these robots act similarly to bees. However, the technology is still in its infancy, and its potential effects on the environment in the long run are still being assessed. Control of Invasive Species Additionally, robots are being used to locate and eradicate invasive species from ecosystems. Native species' survival and environmental equilibrium are both aided by this. Cleaning up the environment Cleanup of polluted areas, such as beaches and oil spills, is aided by robots, reducing the impact of environmental disasters. Robots based on biology Bioinspired robots are made to work in natural settings with little disruption. In conservation efforts, they can carry out activities like exploration, data collection, intervention, and maintenance. Because they are designed to move and sense like animals, these robots are non-invasive and long-lasting conservation tools. The application of robotics to conservation is a promising development in environmental science because it provides novel strategies for preserving biodiversity and improving the health of ecosystems. It is anticipated that the use of these robotic tools in conservation efforts will expand in scope and effectiveness as technology advances, transforming the field.

Chapter 17: Rebuilding Communities After Disasters with Robotic Innovations in Disaster Recovery

Robotic technology is becoming increasingly important in disaster recovery efforts as a result of natural disasters and humanitarian crises. It offers innovative solutions for rapid response, damage assessment, and resilient reconstruction. Robots are changing the way communities recover and rebuild after disasters, from search and rescue to infrastructure repair and removal of debris. One of the most important applications of robotics technology in disaster recovery is in search and rescue operations, where robots equipped with sensors, cameras, and communication systems can navigate hazardous environments and locate survivors trapped in collapsed buildings, rubble, or debris. In this chapter, we will examine the role of robotic innovations in disaster recovery as well as their impact on rebuilding communities and restoring livelihoods.

Ground robots and unmanned aerial vehicles (UAVs) with thermal imaging, LiDAR, and other sensing technologies can survey disaster-affected areas, locate signs of life, and provide vital information to rescue teams, making search and rescue operations more effective and efficient. Also, specialized robots like snake-like robots and unmanned underwater vehicles (UUVs) can get into tight places and underwater environments, which makes it easier for search and rescue teams to work in difficult terrain. Additionally, robotics technology is revolutionizing damage assessment in disaster-affected regions by making it possible to quickly and accurately assess infrastructure damage and environmental dangers. High-resolution cameras and LiDAR sensors can be used in remote sensing drones to look for damaged buildings, bridges, roads, and other critical infrastructure. The drones then provide engineers and planners with detailed 3D maps and digital models that help them figure out how strong the structure is and which repairs should be done first. Additionally, robotics technology is being utilized in debris removal and cleanup operations following disasters, offering efficient and safe solutions for clearing debris, restoring access to critical infrastructure, and preparing sites for reconstruction. Additionally, robotics-enabled sensors and monitoring systems can detect and assess environmental hazards such as chemical spills, radiation leaks, and contamination of the air and water. This enables timely response and mitigation measures to protect public health and safety. Using manipulators and demolition tools, robotic platforms like unmanned ground vehicles (UGVs) and drones can clear debris, excavate sites in hazardous and unstable environments, and remove rubble, speeding up the cleanup process. In addition, autonomous bulldozers and excavators, robotic systems that can move earth and prepare a site, make it possible to quickly rebuild facilities and infrastructure in disaster-affected areas. However, while robotics technology holds great promise for improving disaster recovery efforts, it also raises important ethical, safety, and human impact issues. To ensure that robotics-enabled interventions respect human dignity and promote human well-being, it is necessary to carefully consider ethical concerns about the use of robots in disaster response, such as privacy, consent, and the rights of affected populations. In conclusion, robotics technology is transforming disaster recovery efforts by providing innovative solutions for search and rescue, damage assessment, debris removal, and reconstruction in disaster-affected areas. Additionally, efforts to address safety considerations, such as risk assessment, training, and collaboration protocols, are essential for ensuring the safe and effective deployment of robotics technology in disaster recovery operations. Robots are helping communities recover and rebuild after disasters in a variety of ways, including reducing risk, saving lives, and speeding up recovery and rebuilding efforts. Let us maintain our commitment to promoting the responsible and ethical use of technology and ensuring that robotics-enabled interventions contribute to the construction of resilient communities and the restoration of hope and stability in the face of adversity as we continue to harness the power of robotics in disaster recovery. For robotics technology to have the greatest impact on disaster recovery, it is essential to make efforts to encourage collaboration and coordination among various stakeholders, such as government agencies, humanitarian organizations, technology developers, and local communities. Stakeholders can develop comprehensive and efficient disaster response and recovery strategies by fostering partnerships and knowledge sharing. This will allow them to take advantage of the skills and capabilities of a variety of actors. In addition, efforts to promote innovation and entrepreneurship in the development and deployment of robotics technology for disaster recovery are essential for unlocking new opportunities and scaling up successful initiatives. Additionally, efforts to engage and empower local communities in disaster preparedness and response efforts, such as community-based disaster management initiatives and training programs, are essential for building resilience and promoting self-reliance in the face of disasters. Incentives, grants, and prizes for robotics research and innovation in disaster response and recovery can encourage investment in promising technologies and solutions as well as stimulate creativity. In addition, efforts to address regulatory and policy obstacles in the adoption and deployment of robotics technology for disaster recovery are essential for ensuring that technological solutions are deployed safely, ethically, and effectively. Additionally, initiatives to promote technology transfer and capacity-building in disaster-affected regions can help build local expertise and capacity for using robotics in disaster recovery efforts. Guidelines and regulatory frameworks for the use of robotics technology in disaster response and recovery can help mitigate the risks of unintended consequences and technology misuse, protect the rights and dignity of the affected populations, and ensure compliance with safety standards. In addition, efforts to raise public awareness of and engagement in robotics-enabled disaster recovery are essential for establishing support and momentum for disaster preparedness and response efforts. Additionally, efforts to promote global norms for the responsible and ethical use of robotics technology in disaster recovery are essential. It is possible to raise awareness of disaster risks and encourage proactive measures to mitigate their impact through outreach and education campaigns that highlight the role of robotics technology in disaster response and recovery, showcase innovative solutions and best practices, and involve the public in volunteer and advocacy activities. In conclusion, robotics technology has the potential to transform disaster recovery efforts by providing innovative solutions for search and rescue, damage assessment, debris removal, and reconstruction in disaster-affected areas. Additionally, efforts to promote digital literacy and technological proficiency among diverse audiences can empower individuals to utilize robotics technology for disaster preparedness, response, and recovery in their communities. Robots are helping communities recover and rebuild after disasters in a variety of ways, including reducing risk, saving lives, and speeding up recovery and rebuilding efforts. Let us maintain our commitment to promoting the responsible and ethical use of technology and ensuring that robotics-enabled interventions contribute to the construction of resilient communities and the restoration of hope and stability in the face of adversity as we continue to harness the power of robotics in disaster recovery.

Using technology to rebuild after a disaster

After a disaster, rebuilding efforts rely heavily on technology. The following are some applications of technology: Satellite Data: Satellite imagery may be essential for determining the extent of damage and planning the reconstruction. For instance, plans for redevelopment in Sulawesi, Indonesia, were guided by satellite data following the 2018 earthquake and tsunami. Infrastructure Rebuilds: The idea of "building back better" entails using technology to strengthen infrastructure's resistance to future catastrophes. To lessen the damage caused by floods, this may entail designing roads that soak up water.

- Construction Technology: Rebuilding procedures can be made to run more smoothly and in less time by using automation and other construction technologies.

- Man-made consciousness (simulated intelligence): computer-based intelligence is changing fiasco reactions by anticipating and planning for catastrophes, enhancing reaction endeavors, and working with local area strength.

• Resilience Technologies: New tools are being made to make people more resilient to disasters, like utilities' outage-prediction tools and the use of social media to accurately map out disaster sites. In addition to assisting in the immediate aftermath of a disaster, these technologies also aid in long-term recovery and resilience planning.

Chapter 18: Personal Assistants and Robots: Redefining Daily Life with AI Companions

Personal assistance is changing the way people live their daily lives by incorporating robotics and artificial intelligence (AI) in novel ways that boost productivity, ease of use, and well-being. The way people interact with technology and manage their daily routines is being redefined by robotics technology, which includes virtual assistants, robotic companions, and caregivers. One of the primary applications of robotics and AI in personal assistance is in smart home automation, where interconnected devices and sensors enable seamless control and management of household tasks and systems. In this chapter, we will examine the evolution of robotics and AI in personal assistance and their impact on redefining daily life. Natural language processing and artificial intelligence (AI) algorithms make it possible for smart home assistants to respond to voice commands, manage schedules, and control smart devices like thermostats, lights, appliances, and security systems.

This makes daily routines more convenient and efficient. Additionally, virtual assistants and AI-powered interfaces are revolutionizing the way people interact with information and access services. Robotic vacuum cleaners, lawnmowers, and other autonomous appliances automate household chores, freeing up time and energy for other activities. Natural language commands enable users to access relevant information and services, manage tasks, and organize their schedules with the assistance of virtual assistants like Siri, Alexa, and Google Assistant, which provide personalized assistance and information retrieval. Additionally, robotics technology is being incorporated into wearable devices and personal gadgets, providing personalized assistance and support to individuals in a variety of contexts. Additionally, AI-powered chatbots and virtual agents are being deployed in customer service, healthcare, and other domains to provide personalized assistance and support to users, enhancing accessibility and efficiency in service delivery. Wearable robots like exoskeletons and smart prosthetics make it easier and more independent for people with disabilities or mobility impairments to perform daily tasks on their own. Personal robots and companions with AI algorithms and social interaction capabilities also provide companionship, assistance, and emotional support to those in need, addressing loneliness and social isolation among the elderly and disabled. However, although robotics and AI hold great promise for enhancing personal assistance and enhancing quality of life, they also raise significant concerns regarding privacy, security, and the ethical use of technology. To ensure that individuals' rights and interests are safeguarded, concerns regarding data privacy, surveillance, and the collection and use of personal information by AI-powered systems must be carefully considered. In conclusion, robotics and artificial intelligence are redefining daily life with innovative solutions for personal assistance that offer convenience, efficiency, and support in managing daily tasks and routines. These solutions are essential for promoting the equitable and ethical use of AI in personal assistance. Additionally, efforts to address biases and limitations in AI algorithms, such as fairness, transparency, and accountability, are essential. The way people interact with technology and go about their daily lives is being transformed by robotics technology, which includes wearable robots, virtual assistants, smart home automation, and personal companions. Efforts to promote inclusivity and accessibility in the development and deployment of robotics and AI in personal assistance are essential for ensuring that these technologies benefit all individuals, regardless of age, ability, or background. Let us remain committed to promoting responsible and ethical use of technology and ensuring that robotics-enabled solutions contribute to enhancing the well-being and quality of life for all individuals. Accessibility and usability for people with disabilities or special requirements can be improved by creating user-friendly interfaces, intuitive interaction models, and inclusive features that cater to a variety of preferences and requirements. Additionally, it is essential to address regulatory and policy challenges in the adoption and deployment of robotics and AI in personal assistance to promote responsible and ethical use of technology. Further, it is essential to address regulatory and policy challenges in the adoption and deployment of robotics and AI in personal assistance. To guarantee that individuals' rights and interests are safeguarded, the regulatory frameworks and guidelines that govern the development, deployment, and use of AI-powered systems need to address significant considerations like privacy, security, transparency, and accountability. In addition, efforts to promote education and awareness about robotics and AI in personal assistance are essential for empowering individuals to make informed decisions about technology adoption and use. Additionally, efforts to promote transparency and explainability in AI algorithms and decision-making processes are essential for building trust and confidence among users and stakeholders. Education and training programs that teach people how to use AI-powered systems responsibly and effectively can increase digital literacy and give people the ability to use technology for personal and professional growth. In addition, efforts to promote interdisciplinary collaboration and knowledge exchange among stakeholders, including researchers, developers, policymakers, and end-users, are essential for driving innovation and advancing the field of robotics and AI in personal assistance. In addition, efforts to raise awareness about the potential benefits and risks of robotics and AI in personal assistance, as well as best practices for ethical and responsible use, can foster informed decision-making and promote positive outcomes for individuals and society. In conclusion, robotics and artificial intelligence are reshaping daily life with innovative personal assistance solutions that offer convenience, efficiency, and support in managing daily tasks and routines. Stakeholders can leverage diverse perspectives and expertise to address complex challenges and develop innovative solutions that meet the needs and preferences of individuals in diverse contexts and environments by fostering partnerships and collaboration across sectors and disciplines. The way people interact with technology and go about their daily lives is being transformed by robotics technology, which includes wearable robots, virtual assistants, smart home automation, and personal companions. Let us maintain our commitment to promoting the responsible and ethical use of technology and ensuring that robotics-enabled solutions contribute to enhancing the well-being and quality of life of all individuals as we continue to harness the power of AI and robotics in personal assistance.

Personal Care to Automation of the Home
A growing field that aims to assist individuals, particularly the elderly, in their day-to-day lives is personal care through home automation and robotics. A summary of how personal care and home automation are being transformed by robotics is as follows: Care for the Elderly: Robots are being made to help the elderly live comfortably in their homes. They can assist with daily activities like eating, bathing, dressing, and getting from one location to another. Specialized Systems: Many of these systems aren't humanoid robots but rather specialized machines made to do specific things, like robotic vacuum cleaners. They can be implemented incrementally and are simpler to design and implement. Physical Assistance: Some robots are made to help people get into and out of chairs, beds, and other furniture, follow recipes, fold towels, and give medicine. As a result, independence is preserved and the need for constant human assistance is reduced. Social and Emotional Engagement: Robots also act as social companion for the elderly, engaging them socially and emotionally to help them manage their cognitive decline and slow it down. They can provide therapy and companionship for individuals who are lonely or suffer from conditions related to dementia2.

Automation in Home Care Robot process automation (RPA) employs artificial intelligence and machine learning to automate repetitive home care tasks, which can be advantageous for both patients and caregivers.

> **Future Developments:** With advancements in autonomous vehicles and other technologies that will further integrate robotics into personal assistance and home care, the field is rapidly evolving. Incorporating robotics into home care is not just about convenience; it is also about improving the quality of life for those who require assistance and enabling them to live with greater dignity and independence.

Chapter 19: Research and Development in Robotics: Obstacles and Opportunities

The research and development of robotics is at the forefront of technological innovation and has enormous potential for solving difficult problems and expanding human knowledge and capabilities. However, robotics has unique challenges that must be overcome to realize its full potential, in addition to the opportunities for advancement.

One of the primary challenges in robotics research and development is achieving robustness and reliability in robotic systems, particularly in dynamic and unpredictable environments. In this chapter, we will examine the key challenges and opportunities in robotics research and development, as well as the strategies for navigating the path toward innovation and advancement. It is essential to ensure that robots can operate safely and effectively in a variety of changing conditions as they are increasingly used in real-world applications like manufacturing, healthcare, and disaster response. To improve the robustness and adaptability of robotic systems, innovative solutions in areas like perception, control, and planning are required to address issues like sensor uncertainty, environmental variability, and system complexity. Additionally, scalability and interoperability pose significant difficulties in robotics research and development, particularly as robotics technology becomes increasingly integrated into complex systems and networks. Promoting scalability and adaptability in robotics applications necessitates the creation of modular and standardized interfaces and components that make it possible for robotic systems to seamlessly integrate with existing infrastructure and technologies and interoperate with one another. For enhancing coordination and cooperation among heterogeneous agents, it is essential to address interoperability issues in multi-robot systems and human-robot collaboration. Additionally, addressing the ethical, legal, and societal implications is a significant obstacle in robotics research and development, particularly as robots become increasingly autonomous and pervasive in society. To ensure that robotics technology is developed and used in an ethical, responsible, and beneficial manner for society, it is necessary to carefully consider concerns regarding safety, privacy, accountability, and the impact of robotics on employment and social dynamics. In addition, fostering interdisciplinary collaboration and diversity in robotics research and development is essential for driving innovation and addressing complex challenges from multiple perspectives. Furthermore, efforts to promote transparency, accountability, and public engagement in robotics research and development are essential for building trust and confidence among stakeholders and ensuring that the benefits of robotics technology are equitably distributed. It is possible to foster creativity, cross-pollination of ideas, and holistic approaches to addressing societal challenges with robotics technology by bringing together researchers, engineers, policymakers, ethicists, social scientists, and other stakeholders from a variety of backgrounds and disciplines. In conclusion, robotics research and development present enormous opportunities for addressing complex challenges and advancing human knowledge and capabilities. In addition, efforts to promote diversity and inclusion in the robotics community, including initiatives to support underrepresented groups and foster inclusive research environments, are essential for ensuring that robotics research and development reflect the diverse perspectives and experiences of society. However, major obstacles like robustness, scalability, ethics, and diversity must be addressed before robotics can reach its full potential. We can navigate the path toward innovation and advancement in robotics research and development and unlock the full potential of robotics to benefit society by embracing interdisciplinary collaboration, fostering innovation, and promoting responsible and ethical technology use. Efforts to promote education and training in robotics research and development are essential for nurturing the next generation of robotics researchers and practitioners. We can encourage students to pursue careers in robotics and contribute to field advancements by investing in STEM (Science, Technology, Engineering, and Mathematics) education programs, robotics competitions, and hands-on learning opportunities. Further, fostering collaboration and knowledge sharing among academia, industry, and government is essential for driving innovation and translating research discoveries into practical applications. Additionally, efforts to promote opportunities for lifelong learning and professional development for robotics professionals can guarantee that they remain abreast of the most recent developments and emerging trends in robotics research and technology. Stakeholders can use complementary expertise, resources, and infrastructure to accelerate innovation and address complex robotics research and development challenges by forming partnerships and collaboration frameworks. In addition, efforts to promote open science and open-source development in robotics research and development are essential for advancing knowledge sharing and accelerating progress in the field. Furthermore, efforts to promote technology transfer and commercialization of robotics research can facilitate the translation of scientific discoveries into marketable products and services that benefit society and drive economic growth. Researchers can effectively address major challenges in robotics research and development by adopting open standards, sharing data, code, and resources, and encouraging collaboration across institutional and disciplinary boundaries. In addition, addressing funding and resource constraints is a significant challenge in robotics research and development, particularly for early-stage and high-risk projects. Additionally, efforts to promote transparency and reproducibility in robotics research can enhance the credibility and reliability of research findings and make it easier for the broader research community to replicate and validate results. Stakeholders can support a diverse portfolio of robotics research initiatives and foster innovation in both fundamental science and practical applications by investing in basic research, applied research, and technology development across the entire innovation pipeline. In conclusion, robotics research and development offer enormous opportunities for addressing complex challenges and advancing human knowledge and capabilities. Additionally, efforts to promote public-private partnerships, venture capital investment, and crowdfunding initiatives can leverage additional resources and expertise to support robotics research and development efforts. We can navigate the path toward innovation and advancement in robotics research and development by addressing key challenges such as robustness, scalability, ethics, and diversity, and embracing interdisciplinary collaboration, innovation, and responsible technology use. We can unlock robotics' full potential to benefit society and address the major challenges facing humanity in the 21st century if we work together.

Navigating the Frontier of Robotics Innovation

An exciting journey into a field that combines creativity, engineering, and problem-solving to create intelligent machines capable of performing a variety of tasks is navigating the frontier of robotics innovation. As these machines become more integrated into our day-to-day lives, robotics is about more than just automation; it also involves collaboration, adaptability, and ethical considerations. The following are some significant innovations in robotics: An Overview of the Past: The field of robotics has progressed from early automata to the sophisticated machines of today, with significant milestones such as the development of artificial intelligence and the first industrial robots. Fact versus Reality: In the Indian film "2.0," Chitti's character exemplifies the goals of robotics and how such depictions inspire real-world progress. Applications in Industry: Robotics improve efficiency, precision, and safety in previously difficult or risky tasks, transforming industries. Artificial Intelligence and Robotics: The combination of robotics and artificial intelligence is opening up new realms of learning and adaptability and pushing the boundaries of autonomy and decision-making.

DIY Robotics: There is a thriving community for DIY robotics, and robotics kits encourage a culture of creativity and education among enthusiasts.

Challenges and Ethics: The importance of responsible development is emphasized by the difficulties that rapid development brings, such as job displacement and privacy concerns. Emerging Trends: The dynamic future that lies ahead of us in this field includes emerging trends like soft robotics and swarm robotics. The field of robotics is poised for unprecedented expansion and transformation as we enter a new era, expanding into our homes, hospitals, and even outer space. It's a field that, with these devoted mechanical companions, looks set to shape our future. Accept the journey into innovation, where humans and machines coexist, and push the limits of what was once thought to be impossible.

Chapter 20: The Future of Robotics: Predicting Trends and Designing the World of Tomorrow

The future of robotics holds a lot of promise for shaping the world of tomorrow as we approach a new era marked by technological advancement and innovation. For directing strategic decision-making and preparing for the opportunities and challenges that lie ahead, it is essential to anticipate emerging trends and comprehend the potential impact of robotics on society, economy, and culture.

The convergence of robotics with other emerging technologies, such as artificial intelligence, machine learning, and the Internet of Things (IoT), is one of the key trends shaping the future of robotics. In this final chapter, we will explore the future of robotics and envision the evolution of technology and its transformative impact on our lives and the world around us. We can anticipate a new generation of intelligent and autonomous robots that can learn, adapt, and collaborate in environments that are complex and dynamic as robotics technology becomes increasingly integrated with AI algorithms, data analytics, and connected sensors and devices. Healthcare, transportation, manufacturing, and entertainment are just a few of the industries that will benefit from this convergence of technologies, which will also reshape how we live, work, and interact with technology. In addition, the democratization and decentralization of robotics technology, which will make it possible for a wider range of people to participate in robotics research and development, define the robotics future. Open-source software and hardware, distributed manufacturing, and platforms for collaborative innovation are democratizing access to robotics technology and giving individuals and communities the ability to design, construct, and implement their very own robotic systems for a wide range of uses. The rise of socially and emotionally intelligent robots that can interact with humans in a way that is meaningful and empathetic will also shape the future of robotics. This democratization of robotics technology will drive grassroots innovation, entrepreneurship, and creativity. It will also address the diverse needs and preferences of society. There is a growing demand for robots that can comprehend and respond to human emotions, intentions, and social cues as robots become increasingly integrated into various facets of daily life, such as companionship, education, caregiving, and entertainment. Affective computing, social robotics, and human-robot interaction have made it possible for robots to perceive and interpret human emotions, demonstrate empathy and compassion, and adapt their behavior to social contexts. As a result, interactions between humans and robots are becoming more in-depth and significant. Additionally, as robots become increasingly autonomous and ingrained in society, the future of robotics is characterized by the growing significance of ethical and responsible technology use. To ensure that robotics technology is developed and utilized in ways that are ethical, equitable, and beneficial to society, it is necessary to carefully consider concerns regarding safety, privacy, transparency, accountability, and the impact of robotics on employment and social dynamics. In conclusion, the future of robotics holds immense promise for shaping the world of tomorrow and advancing human progress and well-being. In addition, efforts to promote diversity, inclusion, and social justice in robotics research and development are essential for ensuring that robotics technology reflects the diverse perspectives and experiences of society and addresses the needs and preferences of all individuals. We can harness the transformative power of robotics to address major challenges, foster innovation, and create a more equitable and sustainable future for all by anticipating emerging trends, comprehending the potential impact of robotics on society, and guiding strategic decision-making. Let us embark on this journey into the robotics of the future together, shaping a world in which robots and humans coexist harmoniously, enriching our lives and advancing our shared goals of progress and prosperity. Efforts to promote interdisciplinary collaboration and knowledge exchange will be necessary for driving innovation and addressing complex robotics challenges in the future. Stakeholders can develop holistic solutions to societal challenges and promote the responsible and ethical use of robotics technology by fostering partnerships and collaboration across disciplines like engineering, computer science, neuroscience, psychology, sociology, and ethics. Additionally, efforts to promote robotics education and workforce development will be crucial for preparing the next generation of robotics researchers, engineers, and practitioners. In addition, efforts to engage and empower diverse stakeholders, such as policymakers, industry leaders, academics, and civil society organizations, in dialogue and decision-making processes are essential for ensuring that the benefits of robotics technology are fairly distributed and that the risks and challenges are effectively managed. Stakeholders can encourage students to pursue careers in robotics and contribute to field advancements by investing in STEM education programs, robotics competitions, and hands-on learning experiences. In addition, efforts to address regulatory and policy challenges in the future of robotics will be essential for promoting responsible and ethical use of technology and ensuring that robotics technology benefits society as a whole. Additionally, efforts to promote lifelong learning and professional development opportunities for professionals in robotics can ensure that they remain abreast of the most recent developments and emerging trends in robotics research and technology. To guarantee that robotics technology is developed and utilized in ways that are moral, equitable, and beneficial to society, the regulatory frameworks and guidelines that govern the development, deployment, and use of robotics technology need to address significant aspects like safety, privacy, transparency, accountability, and the social impact. Moreover, endeavors to advance global participation and coordinated efforts on mechanical technology administration and guidelines setting can assist with blending guidelines and advancing worldwide standards for the mindful and moral utilization of mechanical technology.In the end, the fate of mechanical technology holds a colossal commitment to molding the upcoming scene and propelling human advancement and prosperity. We can harness the transformative power of robotics to address major challenges, foster innovation, and create a more equitable and sustainable future for all by anticipating emerging trends, comprehending the potential impact of robotics on society, and guiding strategic decision-making. Let us embark on this journey into the robotics of the future together, shaping a world in which robots and humans coexist harmoniously, enriching our lives and advancing our common goals of progress and prosperity. Efforts to increase public awareness and engagement in the robotics of the future will be necessary for generating support and momentum for robotics research and development initiatives. It is possible to raise public awareness of the transformative impact of robotics on society and inspire public interest and participation through outreach and education campaigns that highlight the potential benefits of robotics technology, showcase innovative applications, and address common misconceptions and concerns. Furthermore, endeavors to advance computerized education and mechanical capability among different crowds can engage people to use mechanical technology innovation for individual and expert development, encouraging a culture of advancement and entrepreneurship.Furthermore, endeavors to address cultural difficulties and advance manageable improvement through mechanical technology innovation will be fundamental for guaranteeing those advanced mechanics propels add to the prosperity and thriving of current and people in the future. Stakeholders can focus on addressing pressing global challenges like poverty, inequality, climate change, and environmental degradation by aligning robotics research and development efforts with the United Nations Sustainable Development Goals (SDGs). Robotic technology can be used as a tool for positive social and environmental impact. Additionally, addressing biases, promoting diversity and inclusion, and mitigating unintended consequences are essential for ensuring that robotics advancements contribute to the construction of a more just, equitable, and sustainable society. world. Additionally, for maximizing the benefits of robotics technology and addressing global challenges, efforts to promote international cooperation and collaboration in the future of robotics will be essential. Stakeholders can use complementary expertise, resources, and infrastructure to accelerate robotics research and development and effectively address shared challenges by fostering partnerships and knowledge exchange among nations and regions. In conclusion, the future of robotics holds enormous promise for shaping the world of tomorrow and advancing human progress and well-being. In addition, efforts to promote technology transfer and capacity-building in developing countries and regions can ensure that robotics technology is accessible to all and affordable to all. We can harness the transformative power of robotics to address great challenges, foster innovation, and create a more equitable and sustainable future for all by promoting sustainable development, addressing societal challenges, fostering international cooperation, and fostering public awareness and engagement. Let us take advantage of the opportunities that lie ahead and work together to shape a future in which robotics technology enhances our lives, strengthens our communities, and advances our common objectives of progress and prosperity. Developing a culture of innovation and entrepreneurship in robotics will be essential to driving economic growth and prosperity. Stakeholders can encourage investment, create jobs, and open up new opportunities for economic development and competitiveness by fostering an ecosystem that supports research and development, technology transfer, and the commercialization of robotics innovations. In addition, it will be essential to address challenges related to privacy, security, and ethical use of robotics technology to build trust and confidence among stakeholders and guarantee that robotics advancements are deployed responsibly and ethically. Additionally, efforts to promote collaboration between academia, industry, and the government, as well as support for startups and small businesses, can accelerate the translation of robotics research into marketable products and services. To ensure that robotics technology is developed and used in a manner that respects individual rights and promotes societal well-being, regulatory frameworks and guidelines that govern the development and deployment of robotics technology need to address important considerations like data privacy, cybersecurity, and algorithmic transparency. They also need to promote principles like fairness, accountability, and transparency. Additionally, addressing disparities in access to robotics technology and opportunities must be prioritized to ensure that the benefits of robotics advancements are equally distributed and that no one is left behind. In addition, efforts to promote public dialogue and engagement on the ethical and social implications of robotics technology can foster a shared understanding of the risks and opportunities associated with robotics advancements. Individuals can be empowered to participate in the robotics revolution and contribute to shaping its future by participating in initiatives that promote digital inclusion, bridge the digital divide, and provide underrepresented groups and marginalized communities with access to education and training in robotics technology. In addition, it is essential to address biases and barriers to participation in robotics research and development as well as diversity and inclusion in the workforce if robotics research and development is to reflect society's diverse perspectives and experiences and maximize talent and creativity. In conclusion, there is a lot of hope for the future of robotics in terms of driving innovation, economic expansion, and social progress. We can harness the transformative power of robotics to create a better future for all by fostering a culture of innovation and entrepreneurship, addressing ethical and social issues, and promoting inclusivity and diversity in robotics research and development. Let us take advantage of the opportunities that lie ahead and collaborate to shape a future in which robotics technology enhances our lives, strengthens our communities, and advances our shared objectives of progress and prosperity. Let us also take advantage of the opportunities that lie ahead.

Envisioning the Next Era of Robotics Integration

Significant advancements in artificial intelligence, machine learning, and automation are expected to characterize the subsequent era of robotics integration as one that will be transformative. The following are some key predictions and trends that are anticipated to shape the robotics landscape in the future: More intelligent AI and machine learning: Robots will get smarter and be able to learn from data and adapt to new situations. Improved Sensory Perceptions: Robots with advanced sensors will be able to interact with their surroundings in greater depth. Smooth Human-Robot Interaction: As robotics become more ingrained in everyday life, the world will become more interconnected. Robotics Democratization: As costs drop, robotic technology will become more affordable for homes, small businesses, and educational institutions. Ethical and employment considerations: To ensure a harmonious coexistence between humans and robots, changes in education, skill development, and social policies will be necessary as robotics advance. These developments will not only enhance the capabilities of robotics currently in use but will also introduce novel applications and solutions to a variety of industries, including healthcare and manufacturing. It is indeed an exciting journey to anticipate and prepare for the future of robotics.

Thank You

www.ingramcontent.com/pod-product-compliance
Lightning Source LLC
Chambersburg PA
CBHW050056230526
45470CB00004B/1550